Also by THOM HARTMANN

The Prophet's Way

Attention Deficit Disorder: A Different Perception

Focus Your Energy

Beyond ADD

Best of the DTP Forum
    (edited with Brad Walrod and Don Arnoldy)

ADD Success Stories

Think Fast!
    (edited with Janie Bowman and Susan Burgess)

Healing ADD

The Greatest Spiritual Secret

Thom Hartmann's Complete Guide to ADHD

Unequal Protection

ADHD Secrets of Success

The Edison Gene

We the People

What Would Jefferson Do?

The Last Hours of Ancient Sunlight

 *The Fate of the World and What
We Can Do Before It's Too Late*

REVISED AND UPDATED

THOM HARTMANN

 THREE RIVERS PRESS • NEW YORK

Published by Three Rivers Press, New York, New York.
Member of the Crown Publishing Group, a division of Random House, Inc.
www.crownpublishing.com

Three Rivers Press and the tugboat design are registered trademarks of Random House, Inc.

Originally published in hardcover in somewhat different form by Mythical Books in 1998, by Harmony Books, a division of Random House, Inc., in 1999, and in paperback by Three Rivers Press, a division of Random House, Inc., in 2000.

Printed in the United States of America

DESIGN BY BARBARA STURMAN

Library of Congress Cataloging-in-Publication Data
Hartmann, Thom
    The last hours of ancient sunlight : The fate of the world and what we can do before it's too late / Thom Hartmann. — Revised and updated ed.
    Includes bibliographical references (p.).
    1. Human ecology. 2. Human ecology—Philosophy. 3. Environmental protection.
GF41.H39   1999
304.2—dc21 99-30268

ISBN 1-4000-5157-6

20 19 18 17 16 15 14 13 12 11

Revised and Updated Edition 2004

To my daughter, Kindra Hartmann,
for all the lessons she has taught us about compassion and love,
and the example of compassion and understanding she lives;

my son, Justin Hartmann,
with respect for his love of books and ideas, and the challenge
he has given to us in his desire and actions taken to change the
world for the better;

and my daughter, Kerith Hartmann,
for the inspiration of her activism and her passionate commitment
to healing both humans and all life.

> *We, the generation that faces the next century, can add
> the ... solemn injunction, "If we don't do the impossible, we
> shall be faced with the unthinkable."*
>
> —PETRA KELLY (1947–1992),
> founder of the German Green Party

Contents

Foreword

BY JOSEPH CHILTON PEARCE,
author of *The Crack in the Cosmic Egg,*
The Magical Child, and *Evolution's End*

To discover brilliant, innovative, creative, and original thinking is one of life's rare privileges. To discover that such a thinker has focused on the most critical of all issues facing our species is encouraging. Because the following work is seminal, offering a viewpoint original yet ancient, I see our situation today in a new light. And, because of the scope and depth of this work, I find it inspiring as well as disturbing. I rather exhausted superlatives in praise of Thom Hartmann's previous book, *The Prophet's Way,* which I rightly called "the most important book I have ever read," since that is what it was to me personally, but now I find this sequel to it calling for equal praise, since the issue here addresses the whole of our species and indeed all species.

I have long puzzled how it is that the heartbreaking and near-terrifying nature of our ecological crisis is treated casually by and large, when not completely neglected or ignored by a generally sleepwalking populace. A planetary crisis embraces everything from the personal and social to worldwide, but in spite of an occasional flurry of lip service and "let's pretends" concerning the avalanche of disasters we are perpetrating, most of our gestures (a bit of recycling, a bit less driving, turning down the heat or AC, sending a check to the Sierra Club) seem to serve only to relieve our guilty conscience or mask our growing feeling of impotence. Nothing much is happening, at any rate, to halt our downward plunge.

Perhaps our tendency to screen out and play ostrich is because the rape of our Earth is simply too huge and awful a predicament, encompassing forces far beyond our personal control (and, it seems with a bit of reflection, everyone else's, since no one is in control). The only close parallel of which I know may be found in that now-receding cloud of atomic-annihilation that hung over our heads for nearly a half-century.

Or, on a more regional (and personal) level, I might mention the destruction by strip-mining, for bituminous coal, of the Appalachian mountains in southwest Virginia and eastern Kentucky, scene of my halcyon and idyllic childhood. The destructiveness of strip-mining, beginning in the post–World War II period, must be seen to be believed. A person can stand atop an untouched ridge and look out and see whole ranges literally leveled, annihilated, gone—ghastly scars gouged into what was once the loveliest of terrains, the southern highlands. Trail of the Lonesome Pine, and all that, now with rivers silted over and ruined, valleys decimated, the land defaced, but huge sums made in the offing— short-term propositions that involved buying up entire judicial systems throughout the region and blocking every conceivable recourse to sanity; earth movers with wheels two stories high (vehicles that had to be constructed on-site) moved mountains without a grain of faith—only petroleum.

Hartmann throws new light on all such issues, showing that our current crisis has a history spanning thousands of years. We but echo forces of thought long since put into motion, woven into our very neural processes of brain. An inherited cultural mind-set building over millennia lies behind our ever increasing acts of violence to our Earth—and one another. The strange anxiety and rage that pervades our modern world is "old karma," of long standing, which we accept as naturally as labeling the color blue as blue or red as red—simply our inheritance, like the father's eating of sour grapes that set the son's teeth on edge generation by generation—an unconscious cycle, but one that can be broken by conscious action, as Hartmann shows.

"Oh, that greens-business" is, however, a common response to the ecological issue, particularly from those successful souls on their way to out-gating Bill Gates. We are so wrapped up in our personal pursuits, simply "getting by," or "making it" in the world, that it's hard to step back and consider the consequences of ordinary daily actions unconsciously accepted as the norm. Or we rationalize, as with a businessman friend of mine who summarized the ecological "dither" with a reference to the enormous size of our planet and our equally enormous insignificance on its surface—as though we microorganisms crawling about on this vast ball could have any significant effect on planetary function!

Thom Hartmann addresses our plight with an admirable simplicity and clarity, backed with an array of research that is sobering indeed. I waved his manuscript about and waxed eloquent over it to friends the other day, but in spite of my enthusiasm, the countenance of my friends changed. "Oh, that . . ." when they found the subject was ecology. They went on with their discussion of things that really mattered—which did not include asides like wounded Earth, a dying ecosystem, a withering human spirit, with massive overpopulation and starvation looming.

I recall my seat companion on a flight from Los Angeles to Washington, D.C., in 1979, a geophysicist in the government's department of geology, just returned from an international meeting of geophysicists in Tokyo. He was hardly an optimist. "Everything our department had predicted over a fifty-year period, and reported to our government," he claimed, "has been borne out to the letter, and yet every report we have made that addresses the gravity of our ecological situation has been as systematically ignored or outright suppressed. We don't have much to contribute to the gross national product, I suppose, and so are dismissed with the comment: 'We need not disturb the public with alarmists like these scientists, who take themselves so seriously, you know.' "

The principal concern of that 1979 international gathering of geo-scientists my traveling companion referred to was the destruction of the equatorial rain-belt forests. The only substantial source of planetary oxygen and absorbent of carbon dioxide around, and a major factor in global rainfall and weather patterns, rainforests were disappearing at the rate of three hundred acres per hour at that time, and according to the consensus of those scientists, a cataclysmic, global error was being made and mounting in tempo.

To find just how that tempo has increased, read on, dear reader, into statistics these 20 years later, if you dare.

Another flying companion-by-chance was a geologist with BP (British Petroleum) in 1991. "We are running out of oil," was his lament. "The end is in sight, yet we continue to burn it," he moaned. "We should conserve petroleum for the vast potential of synthetics and products possible through it. We could develop something like hydrogen fuel instead," he pointed out, "the most common substance in the universe. We're sure it could be adapted to run cars, factories, electric generators, and all that. But burning the last reserves of petroleum," he repeated, "is a disaster of major proportions and vastly illogical."

Were burning our oil reserves the only illogical operation taking place today, our species-wide undercurrent of anxiety might not be so marked, but oil is only one strand in a web of interlocking actions of sheer insanity that Hartmann delineates in the following pages. Why read such stuff, you might ask, if the situation is as grim as the experts claim? Why not eat, drink, and be merry, gathering the few rosebuds we may, since tomorrow . . . ? You should read on because a citizenry informed by superficial three-minute sound bites at best responds superficially. Far more than the spotted owl is at stake. Hartmann's gathering of cold facts is necessary to grasp the full significance of what is happening, and his call to responsible action, an action that any and all of us are capable of making, is a reasonable, practical call to sanity that could, if not turn the situation around, at least plant the seed for a new ecology, a new Earth, a "saving remnant," if you will, when a non-viable and chaotic system finally destroys itself. We first must understand the illness, to understand the prescription for health.

Books on the coming ecological collapse are appearing at an increasing rate, of course, while scientific groups plead with governments, industries, and consumers to heed the signs. All to absolutely no avail. Many of us addressing critical issues of the day preach only to the converted (no one else listens) while the great machine, now grown to the status of "global economy," plunges faster toward a precipice. We recall Rachel Carson's prophetic *Silent Spring* of many years ago, and speak of her work "turning around" the whole issue of chemical poisoning, but realize, in retrospect, that no significant change took place at all, simply cosmetic glosses to ease the outcry and continue unabated what we were doing. Our "environmental protective agencies" continue straining at a few penny-gnats for the cameras, while swallowing vast big-buck camels behind our backs. Carson's contentions, after all, would also have done little to contribute to the gross national product; so mother's milk contains more DDT today than in her time, along with a host of new chemicals she dreamed not of. (We produce an average of nine thousand new chemicals yearly, molecular combinations existing nowhere else in the universe and 90 percent carcinogenic.) And cancer, which ironically laid poor Rachel in her grave, grows near exponentially.

In 1954, really way back there, the British cellular biologists Williamson and Pierce, after 30 years of medical research under a Peckham Foundation grant, spoke of our species as having become a cancer on the living body of our Earth, and predicted that, since Earth and its inhabitants were a single living symbiosis, cancer as a disease of our species would expand into major, epidemic proportions. Their prognoses and recommendations would hardly have contributed to Britain's industrial growth, much less that of the United States, and were totally ignored.

A new viewpoint, a new way of looking, seeing, and presenting the garish facts that should be so self-evident, has long been called for, as a practical way of responding effectively has long been needed. And this is precisely what Thom Hartmann presents here: he makes clear how the impoverishment and decline of the human spirit is at the root of our disease, that these roots are hardly new, but have grown for millennia, and that only a new cultural image of ourselves and life itself will bail us out. We hear ad nauseam of "the new paradigm" emerging in the sciences, but a rediscovery of a very ancient paradigm, that of the sacredness of everyday life, the sanctity of every life-form, of our living Earth, ourselves and each other, alone can turn the tide. And we are not going to be given this image via television or the Internet.

So the last part of this book is not just a call for personal responsibility, a rather vague abstraction, but offers a powerful and articulate "prescription for behavior" even the least of us can follow, both to discover within us this ancient but ever-new image of life, and live that image out. Here is a call for action any and all of us can undertake, to our own personal enrichment, spiritual awakening or renewal, and peace of mind, as well as restoration of our planet.

I can only urge that this remarkable, unusual, unique, and extraordinary book be read by everyone, talked about, promoted by each of us as though our personal life was at stake—which it so vitally is.

Surely we grow tired of "wake-up calls" to action, but this is one we ignore to the peril of ourselves, our children and their children, and this beautiful Earth given to our charge. I thought, on reading Jerry Mander's masterpiece *In the Absence of the Sacred,* which pointed in the same direction Hartmann takes, that I was informed on the ecological issue, but my eyes were opened to new perspectives in *The Last Hours of Ancient Sunlight.*

So read this book! Buy a copy and give it to a friend. Tilt at a few windmills—send a copy to your congressman, senator, Chamber of Commerce, radio station. Insist it become part of every curriculum in every school, college, university. And don't rest on your virtuous laurels with a couple of benevolent gestures (as I tend to do). Keep it up—the deterioration we have helped spin ever faster will not readily abate. Huge forces are in motion. Huge commitment is called for.

On hearing of the slaughter of elephants in Africa, my then-11-year-old daughter, newly possessed of that straightforward and clear logic of the young, and so unable to grasp the murderous irrationality of adults, paced the floor weeping and crying out, "How can they do that? How?"—then turned, pointed to me, and admonished: "And you just sit there!"

What could I say, when I could have done something, but knew not of the means. The following pages, after remarkable insights into the nature of the ills befalling us, offers as even more remarkable means, an outline for concrete action of an unusual and unexpected sort. And even I, here in my seventh decade toward wherever, can do something indeed, as can you. So, as my daughter would say: "Don't just sit there, do something!" Promote this book and live its message. Now. Today is the day, and this is the hour.

Respectfully,
Joseph Chilton Pearce
Faber, Virginia

Introduction

We have had our last chance. If we do not devise some greater and more equitable system, Armageddon will be at our door.

—Douglas MacArthur (1880–1964),
September 2, 1945

In the 24 hours since this time yesterday, over 200,000 acres of rainforest have been destroyed in our world. Fully 13 million tons of toxic chemicals have been released into our environment. Over 45,000 people have died of starvation, 38,000 of them children. And more than 130 plant or animal species have been driven to extinction by the actions of humans. (The last time there was such a rapid loss of species was when the dinosaurs vanished.) And all this just since yesterday.

According to the United Nations' *Pilot Analysis of Global Ecosystems,* half of the world's wetlands have vanished in the past century, half the planet's forests are gone, 80 percent of grasslands and 40 percent of the planet's land surface suffer from soil degeneration, and 70 percent of the planet's major marine fisheries are depleted. They add, "The world's freshwater systems are so degraded that their ability to support human, plant and animal life is greatly in peril."[1]

We in the modern world most often go through our lives concerned with the day-to-day issues of making a living and hanging on to our particular lifestyle. Occasionally, we'll listen a bit more closely, look a bit more deeply, and can find, often with little effort, a cacophony of voices of doom about the future, ranging from the reasonable to the unlikely.

Daily, we can see, hear, or read stories in our media about virulent new bacteria or viruses, increasingly severe weather and killer storms, widespread cancer-causing pollution, alarming threats to our food supply, and those who say that economic collapse and world-wide depression—or perhaps the final battle at Armageddon—are just a few more days or years away.

There are also those who will tell you that everything is just fine, thank you very much, and that there's no problem: the entire human population of the world could fit into an area the size of Florida (although this is not a popular idea among Floridians) and that technology will one day answer all our problems.

Amidst these dueling arguments about what's wrong or right with the world are controversies over what should be done or not done about it.

The problem with most of these arguments and positions is that they overlook four basic realities:

1. Despite the impact of modern technology, the world's dilemmas and dangers are not accidents caused by recent changes. They are the predictable result of the way humanity has been living since the first city-states of the Sumerians were established around seven thousand years ago. Furthermore, they echo repeated cycles such city-states have gone through since some humans decided to move from living in tribes to living in city-states.

2. We (and all other living things) are made up of the food we eat, and the food has sunlight as its sole source of energy. No sun means no living things; abundant sunlight and ample water mean abundant life-forms. We are made of sunlight. How we marshal this most fundamental resource is a reflection of how we see ourselves in relationship to the rest of the natural world.

3. Our problems derive not from our technology, our diet, violence in the media, or any other one thing we do. They arise out of our culture—our view of the world. The reason most solutions offered to the world's crises are impractical is because they arise from the same worldview that caused the problem. As you'll see in this book, recycling won't save the world, birth control won't save the world, and saving what little is left of the rainforests won't save the world. Even if all those good things were fully implemented, our fundamental problem would still remain, and will inevitably be repeated. Even cold fusion and the elimination of the need for oil, with free electricity for

everybody, will not "save the world." Nothing but changing our way of seeing and understanding the world can produce real, meaningful, and lasting change . . . and that change in perspective will naturally lead us to begin to control our populations, save our forests, re-create community, and reduce our wasteful consumption.

4. The solutions I'm proposing in this book are neither new nor radical in the history of the human race. In fact, they represent a view that has sustained and nurtured humanity for tens to hundreds of thousands of years. The indigenous tribes of South America, North America, Africa, Australia, and early Asia did not overpopulate or destroy their world, even though in most cases they had access to far more resources than they used. Neither does the fossil and historical record show that they led rude and desperate lives, as is so often depicted in the media and in the mind of the average person. They lived a sustainable way of life, seeing the sacredness of the world and the presence of the Creator and divinity in all things, and enjoyed far more leisure time than working-class citizens of the industrialized world will ever enjoy. Their consciousness and lifestyle kept their culture and people alive a hundred times longer than the United States has existed, and continues to sustain millions of them worldwide. They have important lessons to teach us—although, as this book will show, we "civilized" peoples are literally exterminating them, and, therefore, risk losing their knowledge as we appropriate their lands, languages, and lives.

When enough people change the way they view things, solutions become evident, often in ways we couldn't even imagine. We have destroyed much of the world because of our culture; we can save much of it by changing our culture. As you read this book, you'll see that in the cultural underpinnings of our ancestors we can find the ancient keys of knowledge to preserve the human race and the planet, without going back to living in caves or huts. In the intentional and voluntary reduction of consumption lies a safer path for humans and the planet. And this book will show you how to achieve a finer quality in your own life as you participate in saving humanity and our world.

* * *

This book is about where the world is headed and what we can do about it. The ending is optimistic, but along the way there are places where what you'll learn is not very good news . . . except that in understanding how things are and how they got that way, we discover tools all around us that are positive and transformational. So, in that context, even the "bad news" is really good news. Guilt and depression are not the goals of this book: I'm writing in the hope of creating positive and lasting change.

The book starts with a portrayal of the state of the world today: population growth, the depletion of our resources, and how we've "fouled our nest" in the process. Vitally important at this stage is a new insight into a factor that few people have realized: the source of the energy that we consume for food and fuel, and the evidence that we truly are in the process of exhausting that source. We'll cover how we got here, and learn why so many people today think that things look just fine, even though they're not.

. People who've worked with me on the book have told me that by the time they've finished reading the first third of the book, their entire understanding of life has changed; they have a new, unsettled, but inescapable view of why things are the way they are, and what it means for the future if we don't do something soon. It's at this point that some travelers turn back or lose their way. But there is hope for the future, even in the face of the problems we see.

The second part of this book explains the "why" of how we got ourselves into this mess. Understanding this "why" is, I believe, the key, and opens the door to new solutions that have already proven viable.

In the last part of the book we'll take the new understandings we've gained and see what we can do with them. If you'll follow along, you'll be left with a sense of realistic, fact-based confidence that if we do the right things, we really can make it to the other side.

Please stay with us. As my friend Gwynne Fisher says, "Hope is the foundation of our maturing." This book is ultimately about hope, and offers concrete solutions for a brighter, more meaningful, and more joyous future.

Thom Hartmann
Montpelier, Vermont

PART I

WE'RE RUNNING OUT OF ANCIENT SUNLIGHT

It all starts with sunlight.

Sunlight pours energy on the Earth, and the energy gets converted from one form to another, in an endless cycle of life, death, and renewal. Some of the sunlight got stored underground, which has provided us with a tremendous "savings account" of energy on which we can draw. Our civilization has developed a vast thirst for this energy, as we've built billions and billions of machines large and small that all depend on fuel and electricity.

But our savings are running low, which will most likely make for some very hard times.

In Part I we'll lay out the situation as a foundation for planning our response. Topics in Part I include:

- The history of sunlight in the human story
- How can things look okay yet be so bad?
- The importance of trees—their three vital roles in a renewable environment, and some alarming statistics on what's happening as we cut them down
- The accelerating rate of species extinction as we alter the world and its climate

Let's start at the beginning, with the fuel source that gave life to this planet millions of years ago: sunlight.

We're Made Out of Sunlight

 The Sun, the hearth of affection and life, pours burning love on the delighted earth.

—ARTHUR RIMBAUD (1854–1891)

In a very real sense, we're all made out of sunlight.

Sunlight radiating heat, visible light, and ultraviolet light is the source of almost all life on Earth. Everything you see alive around you is there because a plant somewhere was able to capture sunlight and store it. All animals live from these plants, whether directly (as with herbivores) or indirectly (as with carnivores, which eat the herbivores). This is true of mammals, insects, birds, amphibians, reptiles, and bacteria . . . everything living. Every life-form on the surface of this planet is here because a plant was able to gather sunlight and store it, and something else was able to eat that plant and take that sunlight energy in to power its body.[1]

In this way, the abundance or lack of abundance of our human food supply was, until the past few hundred years, largely determined by how much sunlight hit the ground. And for all non-human life-forms on the planet, this is still the case—you can see that many of the areas around the equator that are bathed in sunlight are filled with plant and animal life, whereas in the relatively sun-starved polar regions, where sunlight comes in at a thinned-out angle instead of straight-on, there are far fewer living creatures and less diversity among them.

The plant kingdom's method of sunlight storage is quite straightforward. Our atmosphere has billions of tons of carbon in it, most in the form of the gas carbon dioxide, or CO_2. Plants "inhale" this CO_2, and use the energy of sunlight to drive a chemical reaction called photosynthesis in their leaves, which breaks the two atoms of oxygen free from the carbon, producing free carbon (C) and oxygen (O_2). The carbon is then used by the plant to manufacture carbohydrates like cellulose and almost all other plant matter—roots, stems, leaves, fruits, and nuts—and the oxygen is "exhaled" as a waste gas by the plant.

Many people I've met believe that plants are made up of soil—that the tree outside your house, for example, is mostly made from the soil in which it grew. That's a common mistake. That tree is mostly made up of one of the gases in our air (carbon dioxide) and water (hydrogen and oxygen). Trees are solidified air and sunlight.

Here's how it works: plant leaves capture sunlight and use that energy to extract carbon as carbon dioxide from the air, combine it with oxygen and hydrogen from water, to form sugars and other complex carbohydrates (carbohydrates are also made of carbon, hydrogen, and oxygen) such as the cellulose that makes up most of the roots, leaves, and trunk.

When you burn wood, the "sunlight energy" is released in the form of light and heat (from the fire). Most of the carbon in the wood reverses the photosynthesis. The small pile of ash you're left with is all the minerals the huge tree had taken from the soil. Everything else was gas from the air: carbon, hydrogen, and oxygen.

Animals, including humans, cannot create tissues directly from sunlight, water, and air, as plants can. Thus the human population of the planet has always been limited by the amount of readily available plant food (and animals-that-eat-plants food). Because of this, from the dawn of humanity (estimated at 200,000 years ago) until about 40,000 years ago, the world probably never held more than about five million human inhabitants. That's fewer people worldwide than live in Detroit today.

I suspect the reason for this low global census is that people in that time ate only wild-growing food. If sunlight fell on a hundred acres of wild lands producing enough food to feed ten people—through edible fruits, vegetables, seeds, and wild animals that ate the plants—then the population density of that forest would stabilize at that level. Studies of all kinds of animal populations show that mammals—including humans—become less fertile and death rates increase when there is not enough food to sustain a local population. This is nature's population control system for every animal species.

Similarly, people's clothing and shelter back then were made out of plants and animal skins which themselves came to life because of "current sunlight," the sunlight that fell on the ground over the few years of their lives. We used the skins of animals and trees to construct clothing and housing.

Extracting more sunlight—from other animals

Something important happened sometime around 40,000 years ago: humans figured out a way to change the patterns of nature so we could get more sunlight/food than other species did. The human food supply was determined by how many deer or rabbits the local forest could support, or the number of edible plants that could be found or grown in good soil. But in areas where the soil was too poor for farming or forest, supporting only scrub brush and grasses, humans discovered that ruminant (grazing) animals like goats, sheep, and cows could eat those plants that we couldn't, and could therefore convert the daily sunlight captured by the scrub and wild plants on that "useless" land into animal flesh, which we could eat. So if we could increase the number of the ruminant animals through herding and domestication, then we could eat more of the recent sunlight they were consuming as grasses and plants. This provided to our ancestors more usable energy, both as work animals and as food animals. And so domestication and herding were born.

Extracting more sunlight—from the land

About this same time in history, we also figured out that we could replace inedible forests with edible crops. Instead of having a plot of land produce only enough food to feed ten people, that same land could now be worked to feed a hundred. The beginning of agriculture is referred to as the Agricultural Revolution, and it began to gather momentum about 10,000 years ago. Because we had discovered and begun to use these two methods (herding and agriculture) to more efficiently convert the sun's energy into human food, our food supply grew. Following the basic laws of nature, because there was more food, there could be more humans, and the human population started growing faster.

Within a few thousand years of that time we also discovered how to extract mineral ores from the Earth, to smelt pure metals from them, and to build tools from these metals. These tools, such as plows and scythes, made us much more productive farmers, so the period from 8000 B.C. until around the time of Christ saw the human population of the world increase from 5 million people to 250 million people, a number just a bit smaller than the current population of the United States. But we were still only using about one year's worth of sunlight energy per year, and so even though we were eliminating some competing or food species, our impact on the planet remained minimal at worst. We weren't "dipping into our savings" to supply our needs, yet.

Then, as it happened, in the Middle Ages we discovered a new source of sunlight (which had been captured by plants nearly 400 million years ago) that fit in nicely with our new theory that it was acceptable for humans to destroy our competitors for food, to convert all resources of the planet to the production of food for humans: coal, by replacing forests as a source of heat and thus freeing land for agriculture, could be used to increase our production of food.

When ancient sunlight got stored in the Earth

Around 400 million years ago, there was an era that scientists named the Carboniferous Period. Its name derives from the fact that at the beginning of this period there were huge amounts of carbon in the atmosphere in the form of carbon dioxide. Carbon dioxide is a "greenhouse gas," which holds the heat of the Sun against the Earth like the glass of a greenhouse, rather than letting it escape back out into space. During the Carboniferous Period, which lasted 70 million years and extended from 340 to 410 million years ago, there was so much carbon dioxide in the Earth's atmosphere that the temperature of the planet registered much higher than it does today.

The Earth is about 25 percent land and about 75 percent oceans and at that time the entire planet's land mass consisted of one huge continent, which geologists refer to as Pangaea. This continent existed long before the arrival of birds and mammals, even before the dinosaurs, and the only life-forms on the planet were plants, fish, insects, and small reptiles. The high levels of carbon dioxide in the air both trapped sunlight energy as heat and provided copious carbon for the plants to use as raw material, so they grew abundantly. Almost all of Pangaea was covered with a dense mat of vegetation, rising hundreds of feet into the air, creating a thick ground cover of rotting and dead plant matter that became, in some places, hundreds or even thousands of feet deep. The mats of living and dead vegetation became thicker and thicker as this phase continued over 70 million years.

As the plants grew ever more lush, they trapped more and more of the carbon from the atmosphere (converting it into cellulose as leaves, stems, and roots), reducing levels of atmospheric carbon dioxide while retaining that carbon as plant material.

At the same time, the oceans, which cover three-quarters of the Earth's surface, were also home to huge quantities of plant matter, although much of this was of a simpler type, such as single-celled algae and other microscopic plants. These, too, captured the energy of the sun near the surface of the oceans. They used that energy to convert atmospheric carbon dioxide into plant-matter carbon, and then died and settled on the ocean floor.

Approximately 300 million years ago, a massive disaster occurred and created one of the five historical extinctions that have struck our planet. Nobody knows exactly why (a collision with a comet or asteroid is suspected), but a huge explosion of tectonic activity disassembled the continent of Pangaea and irrevocably changed the planetary environment. The Earth's crust was broken open in many places, volcanoes erupted, and continents crumbled and migrated. In those places where the landmasses that were once parts of Pangaea collided with other parts of the former single continent, millions of acres of Earth were covered by mountains or other land. The thick vegetation mat sank underground.

Fifty million years later, the dinosaurs appeared, and another period of relative stability reigned on the Earth and what had become its two major continents, which geologists call Laurasia and Gonwanaland. The Triassic, Jurassic, and Cretaceous Periods (known, together, as the Mesozoic) came to an end 65 million years ago when, according to the most widely accepted scientific view, another meteor or asteroid struck the planet and extinguished the dinosaurs. During the Mesozoic Period, the planet moved into another period of geologic upheaval, and the continents of Laurasia and Gonwanaland broke into smaller parts, creating what we now call Asia, North America, South America, Europe, Australia, Africa, and Antarctica. Mountains were created as continents drifted into each other, and some of the plant matter traveled even deeper into the earth, where it was subjected to great pressure.

Using ancient sunlight

About 900 years ago, humans in Europe and Asia discovered coal below the surface of the Earth and began to burn it. This coal was the surface of most of the ancient mats of vegetation—this 300-million-year-old stored sunlight—and by burning it humans were, for the first time, able to use sunlight energy that had been stored in the distant past. Before this, our ancestors had to maintain a certain acreage of forestland because they needed the wood for heat to survive the cold winters in the northern climates. Forests captured the "current sunlight" energy, and they could liberate that captured sunlight in a fireplace or stove to warm a home, cave, or tipi during the long dark days of winter.

The exploitation of coal, however, reduced their reliance on current sunlight, allowing them to cut more forestland and convert it into cropland, since they no longer were absolutely dependent on the trees for heat. By making more croplands available, they were able to produce more food for more humans, and the population of the world went from 500 million people around the year 1000 to the first billion living humans in 1800.

This represents a critical moment in human history, for this is when our ancestors started living off our planet's sunlight savings.

Because our ancestors could consume sunlight that had been stored by plants millions of years ago, they began for the first time to consume more resources—in food, heat, and other materials—than the daily amount of sunlight falling locally on our planet had historically been able to provide. The planet's human population grew beyond the level that the Earth could sustain if humans were using only local "current sunlight" as an energy and food source.

This meant that if our ancestors' supply of coal had run out, they'd have eventually faced the terrible choice of giving up croplands (risking famine) so they could re-grow forests for heat, or having enough to eat but freezing to death in the winters. (Or, of course, they could have abandoned the colder climates, and packed their population closer together nearer the equator. But the historic movement of people had been away from the equator, a trend encouraged by the availability of fuel.)

We see this same trend today: the availability of a fuel leads to a population that depends on it and will suffer if it is taken away. Had our ancestors run out of coal, Nature would have taken over and limited their population.

Instead, our ancestors discovered another "bank account" they could tap, another reserve of ancient sunlight: the plant matter that hundreds of millions of years ago had sunk to the floor of the oceans, and had then been trapped belowground and compressed into what we refer to as oil.

Oil was first widely used around 1850 in Romania. The real boom began, however, in 1859, when oil was discovered in Titusville, Pennsylvania, in the United States. At that time, the world's population numbered just over one billion people, and the human race was fed both by the current sunlight falling on croplands and their animals' feed crops, and by a substantial amount of ancient sunlight that they dug up by burning coal taken from the Earth in Europe, Asia, and North America.

The discovery of abundant supplies of oil, however, kicked open the door to a truly massive store of ancient sunlight. By using this ancient sunlight locked up with carbon as a heating source and energy source, and by replacing farm animals with tractors, our ancestors dramatically increased their ability to produce food. (Draft animals such as horses and oxen run on "current sunlight": the grass they eat each day, which was grown using recent sunlight. Thus they are limited in the amount of work they can do—whatever they can eat and convert to energy in one day—compared with an oil-fueled tractor that can burn in one day as much sunlight as would be consumed by hundreds of horses.)[2]

More ways to burn ancient sunlight

It turned out that people could use oil for far more than just fuel, so as we moved into the twentieth century, we began "spending" more of our saved-up sunlight.

Oil can be converted to synthetic fabrics (nylon, rayon, polyester), resins for construction of shelter, and plastics (for construction of almost anything, including the keyboard on which this is being typed). Because we could make clothes from oil, we needed less sheep-grazing land and cotton-growing land, thus allowing us to convert even more non-food croplands to food production.

The massive leap in our food supply that began just after the Civil War caused our planet's population to go from just over one billion humans around the time of the discovery of oil to two billion in 1930.

By then, we were beginning to use farm machinery extensively, and the use of oil as a means to increase agricultural production—from running tractors to converting oil into fertilizers to manufacturing pesticides—caused our food production to explode. While it had taken us 200,000 years to produce our first billion people, and 130 years to produce our second billion, the third billion took just 30 years.

In 1960, world human population hit three billion.

But it didn't stop there. We became more efficient at extracting this stored sunlight from oil, distilling it, and making more efficient engines to consume it, and so our food production soared again. As did our population.

It took just fourteen years, from 1960 to 1974, for us to grow to four billion humans worldwide.

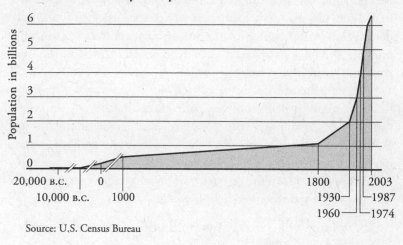

Figure 1. Population growth of modern humans
(*Homo sapiens sapiens*)

Source: U.S. Census Bureau

We added another billion in just thirteen years, hitting five billion in 1987, and our next billion took only twelve years, as the world's human population hit six billion in 1999.

By the fifth billion, in 1987, humans became the most numerous species on Earth in terms of total biomass. Around 1990, we became the most numerous mammalian species on the planet, outnumbering even rats. There is now more human flesh on the planet than there is of any other single species. We now consume more than 40 percent of the world's total "net primary productivity" (NPP), which is the measure of the sum total of food and energy available to all species on Earth. We consume more than 50 percent of the planet's available fresh water. This means that every other species of plant and animal on the planet must now compete against one another for what little we've left.

As is so well documented in Michael Tobias's book *World War III*, we're currently adding a Los Angeles' worth of people to the world every three weeks.[3] In less than a tenth of a percent of the total history of humanity, we've experienced over 90 percent of the total growth of the human population. At the current rate of growth, we would hit 10 billion people in 2030, 20 billion by 2070, and 80 billion by 2150. But nobody expects this rate to continue: there simply isn't enough food that can be produced. Whether what stops it will be famine, plague, natural disasters, or "good science" (such as sudden worldwide availability of and use of birth control) is a source of ongoing debate. But the fact that our current growth rate cannot continue is not in dispute.

We have created this overcrowded world of overtaxed resources by consuming ancient sunlight, converting it into contemporary foods, and consuming those foods to create more human flesh. Without this ancient sunlight, the planet could perhaps sustain between a quarter of a billion and one billion humans—the number it did support prior to the discovery of oil and coal. Without oil and coal, however, the other five billion would starve.

So, how long will our savings hold out? How much fossil fuel do we have left?

And so we begin the twenty-first century standing on a precarious ledge of survival. We are largely dependent on the continued availability of stored sunlight in the form of fossilized ocean plants, the fossil fuel we call oil. And as it happens, the oil is running out.

Since the discovery of oil in Titusville, Pennsylvania, humans have extracted 742 billion barrels of oil from the Earth. Currently, world oil reserves are estimated at about 1,000 billion barrels. To those of us who hope to live another few decades, or have high hopes for our children's and grandchildren's futures, these numbers sound grim. That is, in fact, what the oil industry itself says we can expect, within our children's lifetime.

Oil company executives, however, don't seem to think this is a problem. In an upbeat and optimistic speech presented to the Economic Club of Columbus, Ohio, in 1996, an Ashland Chemical Company executive pointed out that alternatives to oil as an energy source are "simply not cost-effective," but that world oil reserves should last "almost" 45 years, assuming that consumption doesn't increase at all from current-day levels. Citing this as very good news, he concluded his speech by saying that pundits have forecast the end of our oil supplies almost since the first well was drilled by Colonel Drake in 1859. But they've always been wrong in the past. In the happy view of the oil industry, he noted, "it will probably be several decades before the wolf is at the door."

Other experts in the oil industry are less optimistic about the so-called good news that we had in 1996 "almost" a 45-year-supply of oil left in the ground. In fact, the Geneva, Switzerland–based international petroleum-industry consulting firm Petroconsultants points out that North American production of oil peaked in 1974.[4] (By the way, "production" is a nice Orwellian "newspeak" kind of term: we aren't really producing oil, any more than miners "produce" silver. We just pump it out of the ground.)

Back in 1997, in the first edition of this book, I wrote: "World production is expected to peak in the year 2002, when we have consumed over half the world's oil supply. Sometime around this date, they suggest, world-destabilizing price explosions in oil-based products will begin to occur."

They were off by a year: the world's first large-scale oil war was in 2003.

The Petroconsultants study points out that even with consumption dampened by worldwide reductions in oil usage because of increased price (and the probable worldwide depression that this would cause), declining supplies will cause oil production in 2050 to be at levels similar to the 1960s, when the planet only had three billion people on it. But most demographers expect that in 2050 the world population will exceed 10 billion. Imagine: ten billion people alive, but fuel for only three billion. This would leave seven billion people—more than the entire population of the planet today—living on the edge of famine.

Then again, other experts suggest that the oil-industry estimate of 45 years is wildly inflated, meaning the situation is even worse than just described.

Scientist M. King Hubbert first pointed this out in 1956, when he developed the well-known "Hubbert Peak," defining the moment when oil supplies have peaked and then begun a downhill slide. In 1956, he projected a Hubbert Peak for the U.S. in 1970 (he was four years off: the oil crisis was in 1974), and in 1975 predicted a world-wide Hubbert Peak for 1999 or 2000. Although Hubbert died in 1989, his work was carried on by J. Colin Campbell, author of *The Golden Century of Oil: 1950–2050: The Depletion of a Resource,* a book that originated as part of a study of worldwide oil supplies and consumption commissioned by the Norwegian government in 1989.[5] In that book and other sources, Campbell and other scientists point out that oil-producing countries often inflate their estimated oil reserves to qualify for higher OPEC production quotas, and so they can borrow money from the World Bank using their supposed oil supplies as collateral. He and other experts estimate that we're already atop the halfway mark in the world's total oil supply, and that there may be far fewer than 700 billion barrels still in the ground. (An excellent website with both reserves and consumption figures is British Petroleum's at www.bp.com. You can easily divide their reserve figures by their consumption figures.)

It's worth noting that it's unlikely that we'll soon find easily accessible new pools of oil. Most of the world has been digitally "X-rayed" using satellites, seismic data, and computers, in the process of locating 41,000 oil fields; 641,000 exploratory wells have been drilled; and almost all fields that show any promise are well known and factored into the one-trillion-barrel estimate the oil industry uses for world oil reserves.

And, finally, the oil industry's "optimistic" numbers say we have 45 years left at current rates of consumption. But according to data furnished by Petroconsultants (among others), world consumption of oil today is increasing at about 2.8 percent per year. If we were to project that out into the future, our 45-year oil-supply figure drops into the range of just over 30 years.

But that doesn't mean that we'll suck on the straw for 45 years and then it'll suddenly stop: when about half the oil has been removed from an underground oil field, it starts to get much harder (and thus more expensive) to extract the remaining half. The last third to quarter can be excruciatingly expensive to extract—so much so that wells these days that have hit that point are usually just capped because it costs more to extract the oil than it can be sold for, or it's more profitable to ship oil in from the Middle East, even after accounting for the cost of shipping.

At the same time, we'll be adding another billion humans to the planet over the next dozen years, while China, India, Mexico, and the rest of the Third World are industrializing—adding factories and cars, building highways, and constructing oil-fueled power plants—at a growth rate that's faster than both the United States's or Europe's over the past century. So our planet's use is increasing far faster than "current rates of consumption," and our reserves will not last as long as the optimists are suggesting. According to an exhaustive scientific study conducted and released by the British power company Power-Gen and reported worldwide by the Associated Press in September 1997, "Global energy demand is forecast to *double* by 2020" [emphasis added], largely because of the rapid growth of the industrializing nations of Asia, particularly China.

There's obviously a collision coming between our growing population, with its increasing consumption of dwindling supplies of ancient sunlight, and our ability to sustain that population. And even if vast new stores of oil were to be suddenly discovered (there are voices in the oil industry increasingly suggesting this will happen) or if alternative sources of energy such as cold fusion or hydrogen cells became immediately and widely available, their rapid proliferation might actually accelerate the destruction of the planet and the death of billions of humans, in ways that will soon become evident. (On the other hand, there are solutions, as we'll detail later in the book, but they have more to do with our culture than our technology.)

How did it get this way? And what does history tell us about what can be done? We'll discuss these issues and answers in detail in future chapters. But first, let's step back for a moment and look at an important question: if we're headed for trouble, why isn't it obvious?

How Can Things Look So Good Yet Be So Bad?

 Civilization is a conspiracy. . . . Modern life is the silent compact of comfortable folk to keep up pretences.

—JOHN BUCHAN (1875–1940)

There are two ways that things can look fine even when an entire civilization is headed for trouble.

1. Don't "pay as you go"—just live off your "startup capital"

Back in the early 1980s, I was briefly hired as a marketing consultant by a startup company in the computer software business. Four young men had put together about $170,000 in money that they'd earned, saved, or their parents had saved and invested with them. Their plan was to develop and market a new word-processing program that would be better than the then-popular WordStar, thus making them rich.

With their initial $170,000, they rented the second floor of a small office building: they had five private offices, a conference room, and a common area for the secretary they'd hired. They hired a design firm to create a logo, letterhead, and a big sign for the entryway. They leased four Saabs as company cars. They bought oak desks and leather executive chairs. They hired a local florist to install and maintain potted plants, and a local fish store to install and maintain a saltwater aquarium. They paid themselves salaries of $30,000/year each. And they hired me to come in for a few days, and paid me well.

These guys were smart programmers and knew all about computer code. There was no doubt in my mind that they could create a user-friendly and mass-marketable word-processing program. When I first arrived, I was impressed—they and their offices exuded the look and feel of a successful, prosperous business. The young woman in the front office was crisp and efficient, the four founders were well dressed in designer suits and ties, their carpet showed the straight nap lines of the nightly cleaning crew's vacuum cleaners. The heavy-duty copy machine, shredder, postage machine, and office computers all bespoke a thriving business. Top-notch, first-class all the way.

We sat in the conference room, around their oak conference table in comfortable leather chairs, and they confidently told me how they were all going to become multimillionaires. So would anybody who invested with them, they said. Their plan was to concentrate half their attention on raising money from investors, and half their attention on creating and marketing their new product. They'd set a 12-month deadline for themselves to get their new program on the market.

I declined to participate further with them, because I'd seen this unfortunate story played out before with other would-be entrepreneurs, and I was quite certain I knew where it would end.

Six months later, I visited them again, at their request. They now had a staff of 20 people, and the place was humming. Their product would soon be finished, and they had already printed brochures for an upcoming trade show. They were providing employment for the local community and rental revenues for their landlord, had increased their car fleet to six, and raised enough investment capital that they now had a quarter of a million dollars in the bank. They still had not produced or sold anything, but they were going great guns. Things looked great; life was good.

Six months after that second visit, I heard from one of their investors that they had shut down their operation. The four partners had quintupled their salaries, and the company ran out of cash before they could ever bring their product to the market. The company and offices looked bright and sparkling and strong right up to the day the employees were all given 24 hours' notice of termination. The investors lost everything they'd put in because the owners sucked dry their assets before they began to support themselves.

2. The "Ponzi scheme"

A Ponzi scheme is another way in which things go well for everyone, until one day there's a sudden and catastrophic collapse. The story of this all-American entrepreneur is both fascinating and instructive.

In 1917, Charles A. Ponzi was an itinerant housepainter in Florida. World War I was just ending, and the financial systems of Europe were in shambles. Ponzi, sensing an opportunity to capitalize on postwar financial confusion, came upon an idea that would turn him into a millionaire while ruining the lives of thousands.

In late 1919, Ponzi moved to Boston and rented an office on Pie Alley, where he opened a company called the Securities Exchange Company (SEC). His company, he claimed, was incorporated to buy up international postal reply coupons in France and Germany (whose currencies were then vastly devalued) and redeem them in the United States for U.S. currency, thus producing a profit that reflected the difference between the value of the collapsed French and German currencies and the dollar. Such a scheme was, in fact, impossible, but Ponzi made a fortune, and so did his early investors.

Ponzi offered a 50 percent return on investment within only 45 days, and more than 40,000 Bostonians handed over their savings to him. The first few thousand people who invested were richly paid back, complete with the promised interest: Ponzi used money from new investors to fund the paybacks. The first investors told their friends of the quick money they'd made and word spread quickly. At one point Ponzi had a staff of several dozen clerks at Pie Alley who worked long into the nights counting the stacks of cash he was accumulating: over $15 million in less than six months.

During the heyday of his successful enterprise, Charles Ponzi was hailed by a newspaper reporter as the greatest Italian who ever lived.

"You're wrong," he replied with uncharacteristic modesty. "There's Columbus, who discovered America, and Marconi, who discovered radio."

Later unfavorable publicity in the Boston newspaper eventually caused Ponzi's supply of new investors to dry up. Without new money coming in, he couldn't pay out "profits" to his early investors, and so he closed up shop, taking with him the life savings of thousands of unsuspecting investors.

A similar scheme happened in Albania in 1996, nearly causing that country's government to collapse. More than a quarter of all Albanian citizens had put their life savings into one of several huge Ponzi schemes run by local organized crime. Albania's former president, Sali Berisha, said that the government had not stepped in to stop the Ponzi schemes because they had thought such things were normal in a free market and the government didn't want to interfere with the workings of capitalism. Albanians demonstrated and rioted, but to no avail. Their money will never be returned to them.

Our fossil fuel resources: startup capital or Ponzi scheme?

The world is currently living (and growing) by drawing on its "savings account" of energy (sunlight) stored in fossil fuels (oil, coal, gas). Is the world being run like a Ponzi scheme, or like the hopeful software company? I think it's more like the software company, although there are elements of both.

The Earth contains a limited amount of fossil fuels. Although there are different figures for the exact limit, nobody denies that it is there, and we actually have a strong sense of what it is. These fuels are empowering frenetic worldwide activity of much seeming purpose and importance, and that activity is making permanent and irreversible changes in the planetary environment and the human family.

And when the fuel runs out?

Those who earned large paychecks during recent times of prosperity may assume they will have a good chance of survival, and unless there are global epidemics or nuclear war, they may be right. They may even bring a small percentage of the rest of the population along with them, in the way that such things happen in the world of trickle-down economics. The less fortunate may be left holding the food and energy equivalent of what the investors received in Ponzi's scheme and the software company: little or nothing.

When the software guys ran out of running room, they just went out and got another job to support themselves. But when our world economy starts to run out of oil, we can't just close the doors and "go get another energy source." For one thing, through thousands of years of history we've seen that when fuel runs short, wars break out. (More about this in coming chapters.) For another thing, our "other energy sources" aren't yet developed enough.

But there's good news on this point: non-fossil energy sources do exist, and their use is growing. Unfortunately, Pulitzer Prize winner Ross Gelbspan (in his 1997 book *The Heat Is On*) shows that the American oil and coal industries are actively blocking the development of those technologies.[1] As Gelbspan shows clearly, we need to expand our development of the alternatives, so when the oil runs dry there will be somewhere else for our children to turn.

Can we "grow our way out" of it?

In the meantime, we are encouraged by experts and economists to "grow our way out of problems." This solution was first proposed in England in 1954 by R. A. Butler, the British Chancellor of the Exchequer, when he suggested that instead of setting specific growth goals such as building a certain number of homes or new rail lines, the government should simply focus on a steady 3 percent rate of growth. At that rate, he calculated, by 1980 every person living in Britain would be twice as rich.

In fact, the scenario worked out exactly as Butler predicted, according to a 1989 study done by Irish economist Richard Douthwaithe. The problem, however, was that every other index doubled, too. The wealth that went to people at the top of the income pyramid doubled. So did the wealth that went to the poorest, meaning that while a person earning 10 million British pounds a year was now receiving £20 million, a person earning only a thousand pounds a year now earned £2 thousand . . . still stuck in a life of grinding poverty, even though their "standard of life" had slightly improved. In the process, a "social and environmental disaster" occurred, to quote Douthwaithe. Crime increased eightfold, unemployment increased, chronic diseases and mental illness soared, and the divorce rate exploded. All of these were the effects that Douthwaithe had both first predicted and then later chronicled.

Similarly, life has changed for the worse in the United States. On an average day, 100,000 American children carry guns, and 40 are killed or injured by guns, albeit most by accident or as suicides. (A recent bumper sticker says "An armed society is a polite society." One wonders if that sticker's publisher thinks schools today are more polite than a generation ago.) And the dream of a stable family has been replaced by the reality of an army of single-parent children: more than half of all children in the nation.

Around the world, we find that rapid growth is straining almost all nations, and the greatest pain is usually experienced by the individual people and families who do not share in the extreme power and wealth of the society's ruling elite (whether the elite is corporate, governmental, or military).

Technology has, if anything, sped up this process. For example, while at the turn of the century 90 percent of all war casualties were among military personnel, at the end of this century we find that remote-controlled high-tech weaponry (which kills more efficiently and protects *soldiers* from direct combat), and the widespread proliferation of highly efficient weapons, has reversed that proportion: 90 percent of the dead in all wars now are civilians. Over 20 million people have died in wars since World War II, and of those 82 identifiable wars, 79 were internal wars, which hit civilians hardest.

And most of these wars are fought over the control of resources such as forestland, cropland, oil, coal, and minerals.

At a meeting of worldwide central bankers in Hong Kong on September 25, 1997, World Bank president James D. Wolfensohn pointed out that over three billion people—more than half of all humans on the planet, and fully three times the entire human population of the planet in 1800—struggle today to live on less than two dollars a day. "We are living in a time bomb, and unless we take action now, it could explode in our children's faces," Wolfensohn said. About that same time, the Washington-based Population Institute issued a report documenting that 82 nations (more than half the world's countries) have now reached the critical state where they cannot grow enough food nor do they have the resources to pay to import enough food to adequately feed their populations.

The crisis in our oceans

The densest human populations have always been near seacoasts, in part because of easy transport but mostly because the sea has been a historic source of human food. Seafood is not just a traditional meal for much of the world—it's a survival staple, particularly on islands like Japan or Taiwan that don't have enough landmass to grow sufficient food to feed land animals or a vegetarian human population.

In 1994, the U.N. Food and Agriculture Organization released a report that concluded that 70 percent of ocean fish stocks were "fully exploited" (exhausted—all dead) or "overfished" (in rapid decline). The report was challenged by corporations that did large-scale fishing, who brought pressure (and campaign cash) to bear on politicians, so it was largely ignored around the world. But on May 15, 2003, a detailed and thorough analysis of 50 years of data compiled by marine biologists Ransom A. Myers and Boris Worm was published in the prestigious science journal *Nature,* shocking the worlds of both politics and science.

"Analysis of data from five ocean basins reveals a dramatic decline in numbers of large predatory fish (tuna, blue marlins, swordfish and others) since the advent of industrialized fishing," the article's summary says. It goes on to document how "The world's oceans have lost over 90% of large predatory fish, with potentially severe consequences for the ecosystem."

The authors suggest that the U.N. was right when it "argued that three-quarters of the world's fisheries were fished to their sustainable limits or beyond," and pointed out that their data is more accurate than previous information available based "on datasets from commercial fisheries, which can be unreliable." Other evidence is visible from the reports of commercial fishermen, they note. Just after World War II, Japanese fishermen in the deep parts of the Pacific typically caught 10 large fish for every 100 baited "long line" (often a mile or more long) hooks they placed out. Today only one out of a hundred hooks brings back a fish.

Our land is becoming less fertile, plant diseases are spreading and wiping out forests, genes from Genetically Modified Organisms (GMOs) are jumping into the wild with unpredictable consequences, and now we're finding that our fallback—the oceans—are nearly exhausted.

Ancient diseases are re-emerging

But our heavily populated world isn't just stressed by war, poverty, and hunger. Many scientists are alarmed at the potential for epidemic disease created by our high numbers and rapid worldwide mobility. On August 21, 1997, the Associated Press reported that a three-year-old boy in Hong Kong had died a week earlier from a strain of flu that had never before been seen in humans. The flu apparently jumped the species barrier from some type of bird (as apparently happened with the flu that killed 20 million people worldwide in 1918), is deadly, and was identified by laboratories in the United States and Holland as an H4N1 type-A strain, for which there is no vaccine.

The next day, the AP reported that a Michigan man had recently been found infected with a new strain of the ubiquitous *Staphylococcus aureus* bacteria, which is resistant to all known antibiotics, including the newest and most potent antibiotic ever developed, vancomycin. A medical epidemiologist for the Centers for Disease Control in Atlanta, Dr. William Jarvis, said of this first discovery of the killer-staph biological time bomb in the United States, that now "the timer is going off." Three days later, the *Wall Street Journal* reported a second case of vancomycin-resistant staph in the United States, in a New Jersey hospital patient. Since then, it has appeared nationwide.

In Deuteronomy 28:22, reference is made to "consumption," the name most commonly used for tuberculosis (TB) up until just the past 50 years or so. It says: "The Lord shall smite thee with a consumption, and with a fever, and with an inflammation . . . and they shall pursue thee until thou perish." *Consumption?* While some may dismiss this as alarmist—after all, TB is not much in the headlines in American newspapers or on TV these days—consider these facts that reflect the startling new reality of TB in the world:

A recent report prepared by the United States government says: "Among infectious diseases, tuberculosis is the leading killer of adults in the world today and poses a serious challenge to international public health work, according to the World Health Organization (WHO). So great is concern about the worldwide magnitude of the modern TB epidemic that in April 1993 WHO declared tuberculosis to be a 'global emergency'—the first declaration of its kind in WHO history."[2]

The report went on to detail the proportions of the situation: "Someone in the world is newly infected with TB literally with every tick of the clock—one person per second. *Fully one-third of the world's entire population is now infected with the TB bacillus.* [Italics mine: keep in mind that only 5–10 percent of 'infected' people become 'actively sick' and 'contagious.'] . . . TB currently kills more adults each year than AIDS, malaria, and tropical diseases combined. . . ."

One of the problems of TB is that it's so easily spread. As the U.S. Department of Health and Human Services points out: "Like the common cold, and unlike AIDS, the disease [TB] is spread through the air and by relatively casual contact. When infectious people cough, sneeze, talk, or expectorate, the TB bacilli in their lungs are propelled into the air where they can remain suspended for hours and be inhaled by others. Left untreated, a person with active TB will typically infect ten to fifteen other people in the span of a single year."

But won't science save us? Unfortunately, it turns out, modern medical science is what has *caused* much of the problem. While TB is spreading so rapidly, particularly in the developing world where population densities are high (as people try to "grow their way" into a better life), a new and almost incurable form of TB has emerged as a result of the improper use of anti-TB drugs by physicians and hospitals.

Referred to as "multi-drug-resistant (MDR-TB) strains," these forms of the TB bacillus nearly always lead to a painful and agonizing death. As the U.S. Department of Health and Human Services notes: "There is no cure for some multi-drug-resistant strains of TB, and there is concern that they may spread rapidly around the world. While hard data remain scarce, researchers estimate upward of 50 million people are infected with strains of TB that are resistant to at least one of the common anti-TB drugs."

Are you thinking that perhaps this is a Third World problem? An article in the journal *Nature* points out that the disease is already "particularly dangerous" in New York City and Los Angeles, and spreading across the United States.[3] TB travels as fast as a coughing person on an airplane, bus, or train, after all. In the medical journal for surgeons involved in thoracic surgery, *Chest*, the authors point out in the United States an "alarming reversal of the downward trend in incidence of TB," which began around 1984.[4] They state bluntly, "During the past decade, the incidences of HIV infection and TB have ascended to epidemic proportions in several major U.S. cities. At Bellevue Hospital Center in Manhattan, the incidence of MDR-TB has increased sevenfold in 1991 as compared with any of the previous 20 years."

And this, of course, is only one disease. Others of concern include hantavirus, encephalitis, the coronavirus that causes SARS, a repeat of the killer flu of 1918 (the virus that was "carefully" exhumed by researchers in late 1997 so it could be "studied"—they found some victims who had been frozen under layers of permafrost in northern Europe), *Pfisteria piscicida,* which is decimating U.S. East Coast estuaries and waterways, HIV-AIDS, and dozens of others.[5]

There are even problems for meat-eaters in "contagious" proteins (known as prions), the discovery of which earned a scientist the Nobel Prize in 1997.[6] While their discovery in the form of Mad Cow Disease in the United Kingdom in the 1980s created worldwide awareness of that particular manifestation of prions, there are many others, and they are spreading rapidly among both the feed-animal and the worldwide human population.

Even in the face of this and other evidence of the potential dangers and present calamities produced by our explosive growth, those who suggest we may want to consider other slow-growth paths are shouted down as Luddites or environmental extremists or are dismissed as being ignorant of basic economics . . . or even as being "anti-growth," as if what we need is more growth. Notice, though, that the shouting-down is done mainly by those at the top of the growing pyramid of "wealth," which threatens to suffocate our planet.

Things may look good simply because we don't see or hear what's happening

Another reason things may look fine is that, on the whole, Americans are startlingly uninformed about the state of the rest of the world. The American-published *World Almanac and Book of Facts* doesn't even list hunger or famine as categories. It does, however, contain exhaustive listings of American advertisers, American university presidents, American movie stars, members of the United States Congress, and American athletes.

How could it be that in the most prosperous nation on Earth, with by far the greatest media establishment, we're so uninformed? It's an important question.

In today's news it's often the large, multinational corporations who are at the forefront of planetary environmental destruction, but also among the hundred largest corporations in America are the five who own the TV networks that deliver the evening news, which in no small part accounts for why Americans are so ill-informed.

It may pay their bills, but it's not a reliable way for us to know what's going on.

It's also not against the law for news organizations to lie to citizens. On February 14, 2003, the Court of Appeal of the Florida Second District ruled that Fox did not commit a crime when the television corporation ordered television journalist Jane Akre to air what she knew to be false information. Akre and her producer/husband, Steve Wilson, had produced an investigative report that raised questions about how Florida dairies had been secretly injecting rBGH, a genetically engineered hormone manufactured by Monsanto, into cows.

"Every editor has the right to kill a story," said Wilson, "and any honest reporter will tell you that happens from time to time when a news organization's self-interest wins out over the public interest. But when media managers who are not journalists have so little regard for the public trust that they actually order reporters to broadcast false information and slant the truth to curry the favor or avoid the wrath of special interests as happened here, that is the day any responsible reporter has to stand up and say, 'No way!' That is what Jane and I said in our lawsuit."

They lost the suit, however, and the Florida court has now formalized the transformation of the formerly respected profession of journalism into the profit-driven propagandistic corporate tool known now as *infotainment*. Said Wilson after the ruling, "We set out to tell Florida consumers the truth a giant chemical company and a powerful dairy lobby clearly doesn't want them to know. That used to be something investigative reporters won awards for. As we've learned the hard way, it's something you can be fired for these days whenever a news organization places more value on its bottom line than on delivering the news to its viewers honestly."

Although Walter Cronkite and Ralph Nader both testified in Akre's trial about the importance to democracy of honest news reporting, the judge ruled that Fox has the right to order its reporters to lie or face dismissal. Reporters across the nation watched the trial with unease, and when the final ruling came down, *PR Watch* editors Sheldon Rampton and John Stauber said, "Journalists should examine this case and its implications. If the Fox network and Monsanto could destroy the careers of these two seasoned reporters, the same thing could happen to anyone."

In their non-news productions, too, the media show us an idealized reality, not the truth of life in the world. Homeless people, for example, are rarely seen on sitcoms or other shows. The reality, though, is that slowly but perceptibly, parts of the United States are beginning to resemble poverty-pocked Bombay.

Slavery and Freedom

 Slavery is the first step toward civilization. In order to develop it is necessary that things should be much better for some and much worse for others, then those who are better off can develop at the expense of others.

—ALEXANDER HERZEN (1812–1870)

In earlier chapters we've discussed how we're all "made out of sunlight," and that the ability to increase our available sunlight (through fossil fuels) made possible our extraordinary population increases over the past centuries.

Slavery has been another tool of modern civilization, and there are some historians who assert that without slavery the Mesopotamians, Egyptians, Chinese, Greeks, Romans, Ottomans, Europeans, and Americans would not have had anything close to the levels of affluence they enjoyed. (*Science News,* September 20, 1997, mentions "the influential theory that major construction projects and other aspects of complex culture arose only in farming societies that had strict power hierarchies and plenty of slave labor.")

Slavery is another way of taking the sunlight stored in somebody else's body and "harnessing" it for the benefit of the exploiter.

* * *

The earliest history of slavery occurs in the very cradle of Western Civilization: in the Sumerian empire of Mesopotamia, in the Fertile Crescent area around what is now Iraq, five to six thousand years ago. There are also written records of slaves being central to the cultures of Egypt, Persia, Babylonia, and Assyria, as well as extensive mentions (and approvals) of slavery in the Bible (both Old and New Testament). In these societies, the majority of all physical work was done by slaves. As societies expanded and trading networks grew, the demand for slaves increased, leading to the Greek and Roman empires being such heavy users of slaves that, at its height of empire, even the average "commoner" Roman family had at least one household slave, and in the Greek census of 400 B.C., fully a third of Athens' population was slaves.

Aristotle, discussing household management and the essential role slaves played in helping every modern household live the good life, wrote:

> Let us begin by discussing the relation of master and slave. . . .
> For some thinkers hold the function of the master to be a
> definite science. . . . Since property is a part of a household and
> the art of acquiring property a part of household management
> (for without the necessaries even life, as well as the good life, is
> impossible), and since, just as for the particular arts it would be
> necessary for the proper tools to be forthcoming if their work is
> to be accomplished, so also the manager of a household must
> have his tools, and of tools some are lifeless and others living, so
> also an article of property is a tool for the purpose of life, and
> property generally is a collection of tools, and a slave is a live
> article of property. And every assistant is as it were a tool that
> serves for several tools.

By attempting to justify slaveholding as tool-keeping, Aristotle missed the essential point of the contribution that slaves made to Younger Culture civilizations: slaves were not tools, they were *power sources,* kinetic energy, stored energy, expendable energy.

From early civilizations through the current day, slaves have done more than simply provide what Aristotle called the "good life" for their captors. From the African slaves who picked cotton in the American south, to the Russian slaves (the Slavs) imported by the Romans and Portuguese around the year 1000 to work sugar plantations on islands of the Mediterranean, all the way back to Aristotle's household slaves and before, slaves have been a source of power, as in *horsepower* or *energy*. From the slaves of the Roman Empire to disguised forms of slavery such as the serfs of medieval Europe or the wretchedly poor working classes of Victorian England, free or low-cost *backpower, legpower,* and *armpower* were vital fuels for the growth of what we call civilization and industry. One of the most valuable commodities that Columbus found when he blundered into what we now call the Dominican Republic were the natives there—over a period of two decades he shipped thousands of slaves back to Europe, making himself a very wealthy man in the process.

It's interesting to note that the end of slavery in the United States coincided with the advent of widely available oil. U.S. slaves were converting current sunlight (food) into work, which drove the engine of our nation. When coal and then oil became widely and inexpensively available, slaves became less important because we now had machines to replace them with, which were much more efficient users of an ancient source of sunlight, which was more abundant than the current-year's sunlight.

A primary source of slaves for the Romans was warfare: they turned their vanquished "enemies" into slaves. This added to the lure of conquering distant lands: not only could they bring back natural resources such as wood and minerals, but they could bring back slaves as well. Similarly, Europeans shipped more than 12 million African slaves to North and South America during the period between 1500 and 1880, with most going to Brazil and the islands between Florida and Venezuela.

Most people think of the American Plains Indians as warriors on horseback. But the Native Americans of the plains states were pedestrians for 10,000 years, until the introduction of the horse by the Spanish after the failed revolt of Tewa medicine man Pope in 1698. The Native Americans' "Sacred Dog" (their name for the horse) became the pack animal and transportation system of choice among tribes that had previously walked and used dogs to help them hunt. This led to a hundred-year-long golden age among these tribes of unprecedented prosperity and population expansion that came to a terrible and bloody end when Europeans from the east, under Manifest Destiny, decided they wanted the land for themselves.

Nonetheless, the introduction of a new power source, an easier or more efficient way to convert current-sunlight into work—be it slaves or horses or coal- or oil-fired machines—has always dramatically transformed civilizations. Similarly, we see that the loss of these power sources is equally transformational, for they led directly to the decline and destruction of every civilization in our known history, right back to Sumeria.

Survival and prosperity both hinge on how much sunlight energy is under your control.

Glimpsing a Possible Future in Haiti and Other Hot Spots

 The future is made of the same stuff as the present.
—SIMONE WEIL (1909–1943)

Christopher Columbus not only opened the door to a New World, but also set an example for us all.
—GEORGE H.W. BUSH (b. 1924), 1989 speech

If you fly over the island of Hispaniola off Haiti, the island on which Columbus landed, it looks like somebody took a blowtorch and burned away anything green. Even the ocean around the capital of Port-au-Prince is choked for miles with the brown of human sewage and eroded topsoil. From the air, it looks like a lava flow spilling out into the sea.

The history of this small island is, in many ways, a microcosm for what's happening in the whole world.

When Columbus first landed on Hispaniola in 1492, almost the entire island was covered by lush forest. The Taino "Indians" who lived there had an idyllic life prior to Columbus, from the reports left to us by literate members of Columbus's crew, such as Miguel Cuneo.

When Columbus and his crew arrived on their second visit to Hispaniola, however, they took captive about sixteen hundred local villagers who had come out to greet them. Cuneo wrote: "When our ships . . . were to leave for Spain, we gathered . . . one thousand six hundred male and female persons of those Indians, and of these we embarked in our ships on February 17, 1495. . . . For those who remained, we let it be known [to the Spaniards who manned the island's fort] in the vicinity that anyone who wanted to take some of them could do so, to the amount desired, which was done."

Cuneo further notes that he himself took a beautiful teenage Carib girl as his personal slave, a gift from Columbus himself, but that when he attempted to have sex with her, she "resisted with all her strength." So, in his own words, he "thrashed her mercilessly and raped her."

It was a common reward for Columbus's men for him to present them with local women to rape. As he began exporting Taino as slaves to other parts of the world, the sex-slave trade became an important part of the business, as Columbus wrote to a friend in 1500: "A hundred castellanoes [a Spanish coin] are as easily obtained for a woman as for a farm, and it is very general and there are plenty of dealers who go about looking for girls; those from nine to ten [years old] are now in demand."[1]

While Columbus once referred to the Taino Indians as cannibals, there was then and today still is no evidence that this was so. It was apparently a story made up by Columbus—which is to this day still taught in some U.S. schools—to help justify his slaughter and enslavement of the people. He wrote to the Spanish monarchs in 1493: "It is possible, with the name of the Holy Trinity, to sell all the slaves which it is possible to sell. . . . Here there are so many of these slaves, and also brazilwood, that although they are living things they are as good as gold."

However, the Taino turned out not to be particularly good workers in the plantations that the Spaniards and later the French established on Hispaniola: they resented their lands and children being taken, and attempted to fight back against the invaders. Since the Taino were obviously standing in the way of Spain's progress, Columbus sought to impose discipline on them. For even a minor offense, an Indian's nose or ear was cut off, so he could go back to his village to impress the people with the brutality the Spanish were capable of. Columbus attacked them with dogs, skewered them on poles from anus to mouth, and shot them. Eventually, life for the Taino became so unbearable that, as Pedro de Cordoba wrote to King Ferdinand in a 1517 letter, "As a result of the sufferings and hard labor they endured, the Indians choose and have chosen suicide. Occasionally a hundred have committed mass suicide. The women, exhausted by labor, have shunned conception and childbirth. . . . Many, when pregnant, have taken something to abort and have aborted. Others after delivery have killed their children with their own hands, so as not to leave them in such oppressive slavery."

Eventually, Columbus, and later his brother Bartholomew Columbus, whom he left in charge of the island, simply resorted to wiping out the Taino altogether. Prior to Columbus's arrival, most scholars place the population of Haiti/Hispaniola at around 300,000 people. By 1496, it was down to 110,000, according to a census done by Bartholomew Columbus. By 1516, the indigenous population was 12,000, and, according to Las Casas (who was there), by 1542 fewer than 200 natives were alive. By 1555, every single one was dead. (Today not a single Taino is alive: their culture, people, and genes have vanished from the planet.)

As the transplanted population of slaves brought from Africa grew in Haiti, people began cutting the forests to create farmland and to use the trees as firewood for cooking and boiling water. As a result, today trees cover less than 1 percent of Haiti. The denuded land, exposed to rainfall and runoff sped up by the slope of the country's hills, has been so thoroughly eroded that it has mixed with sewage and carried the stain a full four miles out to sea from Port-au-Prince. Millions of people are crowded into the cities, where they provide a ready pool of ultra-cheap labor for multinational corporations, as well as cheap domestic help and inexpensive child and adult prostitutes for the European and American managers of those corporate interests and the occasional tourist.

The legacy of Columbus is that life in Haiti is more than poor; it is desperate. As much as 16 hours a day are spent by the average country-dweller in search of food or firewood, and an equal amount of time is spent by city-dwellers in search of money or edible garbage.

Diseases ranging from cholera to AIDS run rampant through the overcrowded population.

While Haiti is one of the poorest countries in the Western Hemisphere, it is not unique. The Dominican Republic, which shares the island, is moving in the same direction, as is much of the rest of Central and South America.

The Philippines: children hunting for garbage to eat

When I was in the Philippines in 1985, Father Ben Carreon, an activist priest and the author of a popular column for the *Manila Times,* took me to one of that city's huge garbage dumps. The smell was awful, the air thick with insects, as mountains of rotted garbage stretched off into the distance.

We stood in the hot afternoon sun, and Father Ben said, "Look carefully at the piles of garbage."

I squinted in the bright light, looking at the distant piles, and noticed something. "They're moving!" I said.

"No, it's children on them that are moving," he said. "Thousands of them. Their families live all around here, and the children spend their days scavenging for garbage that the family can eat."

Father Ben's response to his discovery years ago that there were armies of children living among the garbage dumps was to begin a scholarship program to put the "garbage dump kids" through grade school and high school. Hundreds have graduated from high school and dozens from college as a result of his efforts. "Still, it's only a drop in the sea," he said to me a few years after we first met. "The task is enormous."

Nepal: walking four hours to find the day's wood

Similar stories are playing out all over the "developing" world. Nepal has given up 30 percent of its forest cover to fuel-wood gathering and subsistence farming in just the past few decades. For the thousands of years that tribal people lived there, elaborate hillside terraces had provided a ready and predictable supply of food for the nation's population. Today, most of those terraces are crumbling under the force of rains that race down Nepal's steep slopes, no longer slowed by forests.

Women in Nepal, as in most developing countries, are the ones primarily responsible for gathering firewood as well as growing or gathering and preparing food. Because of the rapid deforestation of Nepal, studies cited by Dr. Sharon L. Camp of the Population Crisis Committee indicate that Nepalese women have recently had to add between one and four hours to their normal ten-hour workday just to walk to and from the increasingly distant sources of wood. Within the now-visible future, these sources, too, will be exhausted, and Nepal will probably travel the road that Haiti has gone down.

Western Africa: the wood was used up, erosion set in, now it's desert

The western Africa nation of Burkina Faso (formerly known as Upper Volta) is another interesting example. With 18 percent of the country's GNP supplied by foreign aid, Burkina Faso continues to experience a population explosion, with the average woman having 7.2 children. Self-sufficient for tens of thousands of years, the country is now capable of producing only 40 percent of its own food needs. Wood for fuel is being burned almost five times faster than it can regrow, and women spend up to half their waking hours just searching for water. As erosion speeds up and soils become exhausted, the farmers of Burkina Faso have become good customers for the international fertilizer companies, who control a multibillion-dollar annual business. But this is a short-term solution at best, and so the desert has claimed much of the country's land just in the past 40 years.

In 1984, famine killed over a million people across Africa, and Burkina Faso was one of the countries hardest hit. In a 1992 speech, Dr. Camp quoted Burkinabe farmer John Marie Zawadogh, half of whose land had become desert. He said: "In my father's time, millet filled all of the granaries and the soil was deeper than your body before you reached rock. Now we have to buy food in all but the wettest years and the soil is no deeper than my hand. . . . When we were boys, the forest was all around us, too thick to penetrate. Gradually more and more of it was cleared around the compounds, until one clearing met the next and made the great openness you see now."

The United States is no different; it has lost a third of its topsoil since 1950. Yet most people seem unaware that there is a problem here or anywhere else in the world. Why?

We notice rapid changes, not slow ones

In 1976, my wife, Louise, and I bought an 80-acre farm in northern Michigan, thinking that the time might come when it would be necessary for us to grow our own food. We had lived in Detroit when the Arab oil embargo of 1973 happened, followed by the Teamsters' strike over the rise in gas prices and the economic controls that Nixon enforced in an attempt to avoid an economic disaster. For a week or so in 1973, there was little or no food on the shelves of the stores in Detroit, and I remember waiting in line for four hours to buy a five-gallon ration of gasoline. It was clear to us even then that the system was fragile, and that big cities could be death traps if an economic collapse were to happen.

Things improved when the Arabs turned the spigot back on. In 1978, when Louise and I started The New England Salem Children's Village in New Hampshire, we sold the farm in Michigan to help us support ourselves. But I kept remembering that glimpse behind the veil in Detroit, that horrifying look at what a city could be like in just the first few days after the trucks stopped rolling and the pumps ran dry.

A friend who loves seafood once told me that it is possible to cook lobsters slowly. "If you put them in a pot of cold water and then turn on low heat, as it warms up they just go to sleep and then get cooked," he said. "It makes for a lot less thrashing around, as you normally see when you drop a live lobster into a pot of boiling water." The latter method is preferred among lobster aficionados, however, because quick-cooking produces a more flavorful meat, or so I am told.

Not unlike the lobster, we tend not to notice changes in our "water," as long as they happen gradually. For an American, dropping into the "hot pot" of Haiti or Burkina Faso creates a shock of realization: the entire planet is in this same pot, and while there are local spots hotter than others, our "pot" is warming worldwide.

The Death of the Trees

 The development of civilization and industry in general has always shown itself so active in the destruction of forests that everything that has been done for their conservation and production is completely insignificant in comparison.

—KARL MARX (1818–1883), *Das Capital* (1867)

We have already done irreversible (in our lifetime) damage to the soil, water, air, and life-forms of Earth.

More than 75 percent of the topsoil that existed worldwide when Europeans first colonized America is now gone, and substantial damage has been done to the water cycle by cutting our forests. In this chapter we'll explore this subject and learn what it means for our future.

By burning trees, coal, and oil, we're currently pouring over six billion tons of carbon into the atmosphere every year, an explosion compared to the 1.6 billion tons we spat out in 1950. That carbon (most in the form of the gas carbon dioxide) is creating a greenhouse shield that is believed by the United Nations and informed scientists to be causing wild extremes of weather worldwide.

Grain and food production in both America and the rest of the world peaked during the 1980s (and has declined steadily since), leading to both record profits for the agriculture companies and the most widespread hunger and starvation in the history of the planet.

How can it be that our scientific knowledge, which is real and produces tangible benefits, is also leading to a disruption of our existence? The answer is that the tangible results come in isolated specific

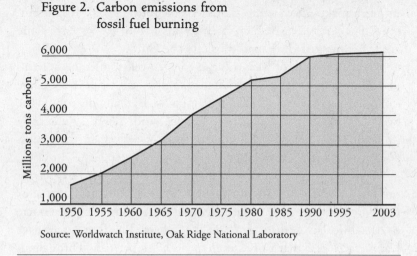

Figure 2. Carbon emissions from fossil fuel burning

Source: Worldwatch Institute, Oak Ridge National Laboratory

arenas, and their gains are accomplished by mortgaging our future: spending one part of the system to benefit another.

Trees

When I was in elementary school, we were taught that the oceans and the forests were the chief sources of oxygen for the planet. It turns out that, at least for those animals that breathe air, this is only partially correct. The oceans account for less than 8 percent of the atmosphere's oxygen, and that is dropping rapidly: there are now millions of acres of ocean that are dying from the dumping of toxic wastes or changes in water temperature and therefore have become net *consumers* of oxygen.

For example, at a January 1999 meeting of the American Association for the Advancement of Science, researchers reported that the seven-thousand-square-mile "dead zone" in the Gulf of Mexico has doubled in size since 1992, leaving a huge area now devoid of fish, shrimp, and almost every other form of life except certain bacteria that prefer low-oxygen environments. The cause, according to Purdue University professor Otto Doering, is related to the 6.5 million metric tons of nitrogen dumped as fertilizer on U.S. agricultural land every year by farmers practicing intensive agricultural methods. This nitrogen makes its way into thousands of waterways that drain into the Mississippi River (which itself drains about 40 percent of the entire continental U.S.), and thus into the Gulf.

While the Gulf of Mexico dead zone is well studied because it's just off the coast of the United States, similar oceanic dead zones are exploding around the world, threatening fisheries and disturbing the overall ecosystem of the planet's oceans. And they're significantly decreasing the already small contribution of oxygen to the atmosphere traditionally provided by the oceans.

So trees, it turns out, are *the* major source of recycled oxygen for the atmosphere. They are our planet's lungs. A fully grown pine or hardwood tree has a leaf surface area that can run from a quarter-acre to over three acres, depending on the species. Rainforest trees have leaf surface areas that run as high as 40 acres per tree. Throughout this enormous surface area, sunlight is used as an energy source to drive the conversion of carbon dioxide into oxygen and plant matter (using the "C," which is carbon). Trees literally breathe in the CO_2 through that enormous leaf area after we exhale it as biological waste, and they exhale oxygen as their own waste. Without trees, our atmosphere would most likely become toxic to us, and because rainforest trees have such a massively larger leaf area than our common trees, the rainforests of the world provide much of the oxygen that you are breathing as you read this page.

While this is common knowledge, it's really among the least important functions that trees play: other details about trees' role in our survival are less well known.

The root system "water pump"

A rainforest tree will draw three million gallons of water up through its roots and release it into the atmosphere as water vapor during its lifetime. While it may seem that this would deplete the soil of water, actually the reverse is true: trees draw water *into* the soil, the first step in a complex cycle that prevents land from becoming desert.

Without forestland pumping millions of tons of water into an area's atmosphere, there's little moisture released into the air to condense into clouds and then fall again as rain. The result is that just downwind of the place that was once forest but is now denuded, the rains no longer fall and a process called *desertification* begins. This has happened over much of north and eastern Africa, leading to massive famines as the rains stop, crops fail, the topsoil is blown away, and what is left is desert.

(Most rainfall on non-forestland is either absorbed and becomes surface ground water or transported along culverts, ditches, sewers, streams, and/or rivers, eventually reaching the ocean. On our continental landmasses, only *trees* effectively cycle large quantities of water back up into the atmosphere. For comparison, think about the evaporation from a 40-acre lake. That may seem like a lot of water to be evaporating into the atmosphere, but that 40 acres is also the evaporative leaf surface of a *single* large tree.)

As of this writing, over 1,500 acres of land are becoming desert worldwide every hour, largely because of the destruction of upwind forests. The total amount of rainforest left on the planet is about the size of the continental United States, and every year, an area the size of Florida is cut down and permanently destroyed.

Reseeded saplings can't pull the water down

The timber industry's ads that show loggers planting seedlings after stripping trees from a forest are utterly misleading with regard to the water cycle. They may well be replacing trees, but they're creating a decades-long gap in the water cycle.

Another problem is that they're setting up an ecological disaster by planting the same species throughout a deforested area. When an entire forest is all made of the same species of tree, and they're all the same age, it becomes an irresible treat for tree-eating caterpillars, beetles, and fungi, as we've seen in numerous forests in North America and Europe.

Taking thousands of tons of biomass (fully grown trees and habitat) out of a forest and replacing it with saplings that weigh a few ounces will do little for the downwind areas that need the atmospheric moisture to produce rainfall. Even by the time the trees regenerate, the ecological diversity and natural fauna and flora of the region have been decimated as the diversity of numerous plant species are replaced by the single-species seedlings used by the loggers. But it's not just the timber companies who are responsible for the destruction of the planet's forests.

Trees for beef: slashing rainforests so Americans can have a 99-cent burger

According to a 1996 report by the Consultative Group on International Agricultural Research, funded by the World Bank and the United Nations, *72 acres of rainforest are destroyed every minute,* mostly by impoverished people working for multinational corporations, who are cutting and burning the forest to create agricultural or pasturelands *to grow beef for export to the United States.*

This 38-million-acres-per-year loss will wipe out the entire world's rainforests in our children's lifetimes if it continues at its current pace. The end, literally, is within sight.

A spokesman for the World Bank said the study pointed out that poverty and overpopulation are the primary factors leading to the destruction of these forests, which are so essential to maintaining human life on the planet. He conveniently overlooked the role of huge agricultural corporations.

Recently, a friend of my son complained to me that one of the giant fast-food hamburger chains was responsible for the destruction of many of the rainforests in the Americas. I didn't understand what he meant: the assumption I'd always had was that the rainforests were cut by timber companies eager to sell rare woods to Japan and Scandinavia for manufacture of furniture and specialty items. If the fast-food chains were killing off the rainforests, I thought, it must be because they were buying cheap wood for paper to wrap their burgers in, or that their plastic packaging was somehow damaging to the rainforests.

It turns out, however, that I shared a common misconception, and one that I'm sure the American fast-food industry is probably quite happy keeping intact.

While these rainforests that have taken centuries to grow are often logged and the wood is sold, they're just as often simply burned and not reseeded, particularly if it's in places where it's inconvenient to take the wood to market. The "free" wood is usually only an added bonus, a quick buck for a peasant farmer to use to buy some breeder cattle.

The most common reason why people are destroying most of the South and Central American rainforests is corporate greed: the American meat habit has provided an economic boom to multinational corporate ranchers, and it is the primary reason behind the destruction of the tropical rainforests of the Americas. Poor farmers and factory farmers alike engage in slash-and-burn agriculture, cutting ancient forests to plant a single crop: grass for cattle.

As John Robbins points out in his book *Diet for a New America,* "The United States imports two hundred million pounds of beef every year from El Salvador, Guatemala, Nicaragua, Honduras, Costa Rica, and Panama—while the average citizen in those countries eats less meat each year than the average American house cat."[1]

This deforestation of Latin America for burgers is particularly distressing when you consider that this very fragile area contains 58 percent of the entire planet's rainforests (19 percent are in Africa and 23 percent in Oceania and Southeast Asia).

Deforesting removes roots, affecting groundwater and the water cycle

Another problem relating to deforestation is the loss of drinkable groundwater.

Drinkable water falls from the skies as rain and soaks into the ground.

At deeper levels, the water has often acquired (from the soil) high concentrations of dissolved minerals, particularly salts. Trees reach deep down into the earth and draw up moisture from just above this salty water and pump it up into the atmosphere, using the minerals to harden the wood of the tree. This removal of water from the soil creates a downward draw, into the soil, for the fresh water raining down from above. This circulation keeps the soil healthy.

When forests are cut, however, the more saline subterranean water begins to creep upward, infiltrating into higher and higher levels of soil. When this salty water hits a level of a few yards below the surface, the remaining trees become immune damaged, just like an AIDS patient, vulnerable to parasitic infections. We see the result of this in beetle infestations and fungal infections such as "rust," which are wiping out trees around the world.

People often think that beetle, caterpillar, moth, and fungus infections are external agents that cause forests to die, and react to them with mass sprayings of insecticides or fungicides, or by shrugging their shoulders and saying nothing can be done. But in a healthy forest such infestations are rare, just as in a healthy human opportunistic infections are rare. One reason why even multispecied, varied-aged tracts of forest in Europe and the United States are dying from these conditions is because they've already been weakened by humans pumping out much of the surface water, pouring down acid rain on them, and destroying surrounding forests.

In Europe the percentage of land that is forest is reduced to 27 percent. In Asia it's 19 percent. In North America (including the vasts forests of Canada), it's at 25 percent. The worldwide replacement of forests with pastureland for cows has become so pervasive that wood-poor England is now, in some communities, using charcoal made from burned cow bones instead of the traditional wood charcoal to filter city water supplies. Reacting to protests from vegetarians in Yorkshire, England, the Yorkshire Water company pointed out that the bones were imported from India, because the company couldn't afford the cost of wood-made charcoal and, the Associated Press quoted an official as saying, "We can't undertake to supply water which meets individual dietary needs."[2] As of 1997, cow-bone charcoal, cheaper than wood charcoal even after including the cost of shipping it from India, is now being used in ten water-treatment plants, and the company plans to add it to six more in the coming months.

When the salty water continues higher and reaches a foot or two below the surface, crops begin to die. And when it hits the surface, the soil becomes incapable of sustaining vegetation and desertification sets in.

To deal with this growing soil salinity crisis, farmers from California to Europe to Australia have begun installing deep-water pumps to remove the salt-contaminted water that the trees would have once drawn down deep below the surface. While this works as a short-term solution, over the long term it only makes the problem worse because that undesirable water is not being cycled back up into the atmosphere as it would be by a tree, but instead is dumped into waterways where it poisons them on its way to the sea. The result is further downwind desertification as well as the poisoning of rivers and lakes.

This mineral and salt contamination of groundwater is also a crisis for thirsty humans. In many parts of the world, city drinking water is so brackish it is dangerous. Most major U.S. and European cities have water that is, at best, unpalatable. Dissolved salt levels of 1300 ppm (parts per million) are the point where people begin to become sick and dizzy from drinking water: in many cities levels now exceed 1000 ppm.

The loss of trees means not only the loss of current topsoil because of salination and desertification, but also the loss of future soils. The roots of most plants anchor only into the topsoil, using it for mechanical support and as a medium from which to derive nutrients and water. Trees, however, have deep roots that break up lower levels of rock, slowly bringing them to the surface, and shallow roots that break up surface rock. They also draw minerals up into the tree itself to help make the plant matter. When the leaves are shed, they form an essential component of soil.

The result of this action by the roots of trees is the formation of new topsoil. It takes, on average, about 400 years for a forest to create a foot of topsoil that is capable of sustaining crops. Without a forest there is almost no topsoil being created at all. (Some sand is formed through air and water erosion of rock, but that is not soil.) This also shows how "slash-and-burn" agriculture, where a few feet of topsoil are exposed by burning a forest and then used up by agriculture over a few short years, is so shortsighted.

Given that without soil we can have no crops, it would seem that we'd be concerned about both the loss of our soil-creating trees and the loss of our current soil itself. Instead, over 300 tons of topsoil are lost worldwide *every minute* as governments and the agricultural corporations that produce most of America's crops look the other way.

Because of rising average temperatures from global warming, the life cycle of the bark beetle in Alaska has been cut from two years to one for reproduction. This has led to a near doubling of the population of bark beetles, which have devastated several million acres of Alaskan forests.

Forests are imperiled worldwide.

*   *   *

Hardly anything illustrates the rich, complex, interdependent nature of our environment as well as trees do, but they continue to be cut and burned. The result aggravates our situation in these last hours of ancient sunlight: we have less oxygen-releasing leaf surface, less circulation in the water cycle, and increased desertification, while at the same time the burning puts more carbon into the atmosphere. These facts make it appear that humans (at least the humans who control such matters) have no concept of their role in the ecosystem. But the domination is now weakening us in another way, too: the same extermination mentality that killed off the Taino (and any other population that interfered with the dominators) is killing off species at an absolutely unprecedented rate, resulting in another change that cannot be quickly undone.

Extinctions: Diversity Supports Survival

 The nation that destroys its soil destroys itself.
—FRANKLIN D. ROOSEVELT (1882–1945)

Modern humans first appeared around 200,000 years ago. (Some estimates range from 400,000 to 40,000 years, but 200,000 is the most commonly accepted figure, based on the fossil record.) Up to the birth of Christ—the first 198,000 years—the world population grew to about 250 million people.

But even those first quarter-billion humans had significant impact on the species of the world. For example, in North America we no longer see many animals that were part of the ecosystem 20,000 years ago (unless we look in the La Brea Tar Pits and fossil digs). Gone are the giant woolly mammoth, saber-toothed tigers, elephants, giant bears and sloths, the lumbering glyptodons (a very large armadillo-like animal), wild ancestors of horses, and camels, among others.

Around 10,000 to 12,000 years ago, these animals and 57 other major species of large mammals vanished from the Americas: an extinction that occurred, on the planet's timescale, in the blink of an eye.

But why?

The popular theory is that they died as the result of a climatic change brought about by the end of the Ice Age, around 12,000 years ago. But recent research reported in detail by Richard Leakey has shown some significant holes in that theory.

For example, similar mass extinctions occurred in the Pacific islands (including Hawaii), Australia, and New Zealand. Killed off to the point of extinction in a thousand years or less were hundreds of large ground animals, including flightless birds, tapirs, rhino-like animals, a giant lizard bigger than the Komodo dragon, an elephant-sized mammal, and giant ground sloths.

But the extinctions in Australia, New Zealand, and the other Pacific islands occurred at different times from those in the Americas, even though the end of the Ice Age affected all parts of the world equally. Why?

Paleontologist Paul Martin of the University of Arizona points out that while changes in the weather in these different places didn't coincide with the mass extinctions of large ground animals, another event did—the sudden appearance on the scene of the most deadly and wanton predator the Earth has ever known: man.

"Clovis people" is the name given by paleontologists and archaeologists to the humans who crossed the Bering land bridge from Asia and arrived in the Americas 11,500 years ago. Within just 350 years, according to Martin, Clovis people had reached the Gulf of Mexico and their numbers had increased to just over a half-million people. By 10,500 years ago they'd reached all the way to the southernmost tip of South America.

Along the way, they left souvenirs for paleontologists to discover: arrowheads and spear points, scattered among the fossilized remains of many of the now-extinct species. (Their spear points were first found and identified in Clovis, New Mexico.)

As Leakey points out graphically in *The Sixth Extinction*, the extinctions of animals in Australia (about 20,000 years ago), North America (about 10,000 years ago), and Madagascar and New Zealand (about 1,000 years ago) all coincided perfectly not with climatic changes, but with the arrival of humans in those places.[1] People who've proposed this "Pleistocene Overkill" hypothesis in the past have been countered with the question, "Well, if all those animals were killed by humans, how did the bison and buffalo, four types of kangaroo, bears, and other species survive?"

Leakey proposes an elegant answer that Darwin would have felt right at home with. He suggests that those animals hunted into extinction were the ones who had few natural predators and were therefore unafraid of this new, small, hairless animal. They had no idea that humans could be so deadly, and were wiped out before they had a chance to breed generations of human-wary progeny. The animals that survived the onslaught of man were the ones instinctively wary of *everything* in their environment—including man.

So we can see that even this very early and relatively small worldwide human population had a substantial impact on the planet, one that in all probability led to a significant extinction of species. But now, with the added power of fossil fuels, our population and our impact have amplified to the point where we're endangering continent-wide and planet-wide ecosystems.

Diversity supports survival, and we're losing it

We are facing an implosion in the form of loss of diversity, from ecological to economic systems.

In mid-1996, a power outage struck many of the western states, leaving millions without electricity for most of a day. Hospitals had to go to emergency power, people were trapped in elevators in hundreds of cities, people sweated without air-conditioning as temperatures in the region exceeded 100° Fahrenheit. It turns out that the crash of the western power grid was the result of some trees in Oregon not being properly pruned. On a particularly hot day, some high-tension lines began to stretch and sag, as metal does when it's heated. They sagged into the trees and shorted out the lines, blowing out part of the Northwest power grid. Because this grid was supplying surplus power to the California/Nevada area, the loss of the surplus put an overload strain on that system and brought it crashing down. Every time they tried to start it back up, it would crash back down, until engineers found the fried trees and restored the Oregon system. And in 2003 there were major outages in both the northeastern U.S. and the whole of Italy.

This domino effect shows how a small change in one part of a complex system can produce huge changes elsewhere. It's been long known to electrical engineers: it's how transistors are capable of amplifying the feeble current from a phonograph needle into earsplitting sound from a speaker. But most people don't realize how fragile it makes human and ecological systems.

When systems are small, local, and widely scattered, they're relatively immune to failure

When people heated with wood, used sunlight and candles for light, and grew and hunted their own food locally, a problem in one part of the country had little effect on another part of the country.

Similarly, when people grew and ate a variety of foods, they weren't affected by the destruction of a single species. But in Ireland, when they allowed potatoes to become a primary staple, a failure of that crop led to widespread starvation in 1846.

America (and most of the rest of the world) has been on a binge of centralization of services and products. While there are over 15,000 known edible plants that grow in North America alone, most Americans eat fewer than 30 plants in the average year and fewer than 50 in their lifetime. Huge ranges of cropland are planted with the same (often hybrid) crops, a massive petri dish just waiting for an infectious agent.

Most of our food production is provided by a small number of huge companies; these firms hold our survival in their hands.[2] Indeed, they are so aware of this fact that many hybrids are intentionally bred to produce sterile seeds so that farmers have to keep buying new seeds every year. (If you find this hard to believe, consider this: in the past decades several farmers have been *prosecuted for theft* by seed companies for keeping some of their own crops' seeds to replant the next year, and others for keeping seed wind-pollinated by genetically modified crops on other farmers' fields.)

The normal, or "background," rate of species loss is one species every four years, according to Richard Leakey. That background rate of loss held constant for over 300 million years—the planet losing on average 25 species every century, or 250 species every thousand years—until this century. At the current rate of human destruction of planetary ecosystems, the Earth has lost nearly one-quarter of all species of plant and animal life that were present when man first appeared. This loss has happened largely in the last one hundred years.

Because of the presence of over five billion humans on the planet, we are losing species at a rate of 17,000 to 100,000 a year (depending on whose numbers you use): a worldwide implosion of plant and animal life that has only been equaled five times in the past five billion years (the last time being the death of the dinosaurs).

This, says Leakey, qualifies as a mass extinction, and has thrown the entire balance of nature out of kilter. And, he points out bluntly, the animal at the top of the pyramid—which caused the extinction of those species that supported and fed it—will itself soon face a mass extinction if things don't change radically and rapidly.

Social diversity, too, is suffering

The predatory way we're wiping out other species is both reflected in and partially caused by the obsession in our culture to accumulate wealth, often with no regard to that accumulation's consequences to the ecosystem or to other humans. If taking the resources of other species is acceptable, why not take the resources of other humans, too? If exploiting other species is a good thing, why not exploit other humans, too? Consider these statistics from the United Nations Development Program:

• The difference in wealth between the world's richest and poorest people slowly grew over the first two-thirds of this century. But in 1960, an explosion began: between then and 1989, the distance between rich and poor doubled.

• As of 2003, the richest 20 percent of the world's population controlled over 87 percent of the world's wealth, whereas the poorest fifth of the world had access to only 1.4 percent. That's a ratio of 60:1. We approached such an imbalance just before the stock market crash of 1929 (around 40:1), but other than that time, such an imbalance has never been seen in a "democratic" economy that survived, although it's common in ones that have flipped from democracy to dictatorship or anarchy, such as numerous African nations, pre–World War II Germany, pre–Revolutionary France, etc.

• The Northern Hemisphere countries (North America, Europe, northern Asia) contain only 25 percent of the world's population, but they consume over 70 percent of the world's total energy stores, eat more than 60 percent of its food, and consume over 85 percent of its wood.

• While we're accumulating wealth and consuming resources at this incredible rate, thousands of people die from hunger worldwide every hour.

The consolidation of power and wealth in the hands of a very few rich individuals and multinational corporations has made some businessmen and politicians rich, but it's also aggregating and wiping out resources: we're directly competing with every other form of life on the planet. As long as there was "more out there" to exploit, growth was possible. Now, as we approach the closed limits of our planet's capacities to generate food and process our wastes, "sustainable growth" has to be re-examined for the oxymoron it is. (This is brilliantly laid out in World Bank economist and University of Maryland professor Herman Daly's book *Beyond Growth*.[3])

And even if nature doesn't kill us off, it seems that we're bent on doing it to ourselves. Pesticide use in the United States is up over 3,000 percent since World War II, yet more pesticides haven't meant fewer crops lost to insects. To the contrary, we're losing 20 percent more of our crops to insects today than in 1945, but because of increasing insect resistance to pesticides and mechanized farming techniques, the pesticide industry has economically addicted many farmers to their product. Harmless species *are* disappearing, but *no* harmful insect species has been eradicated. While the insects—who can evolve through hundreds to millions of generations during the period of a single human generation—are becoming immune to our pesticides, we are not. This leaves us vulnerable to the poisons we, ourselves, manufactured to kill off other species.

For example, in September 1997, the *New York Times* featured a story by reporter John H. Cushman Jr. titled "Cancer in Kids Increases: New Toxins Suspected." The story chronicled how the rate of cancer among children in the United States has skyrocketed since the 1970s—when we were using far less than half of the agricultural chemicals we are now—to the point now where the odds are 1 in 600 of a child born today getting cancer before the age of 10. Childhood cancer has become the second-leading cause of childhood deaths (after accidents), and is now the most common fatal childhood disease, accounting for 10 percent of all childhood deaths. Since 1973, for example, rates of acute lymphoblastic leukemia increased 27 percent in boys and girls, and brain cancer is up 40 percent over the same period.

Ninety-nine percent of all U.S. mothers' milk *today* contains detectable levels of DDT.

In 1950 it was found that half of 1 percent (0.5 percent) of U.S. male college students were infertile. In 1978, a study found that number had skyrocketed by 25 percent, and in the past 32 years the average American male's sperm production has dropped by 30 percent. Some researchers attribute this to the accumulation of chlorinated hydrocarbon pesticides (which often are intended to render insects sterile), while others speculate that some plastics used in food packaging mimic the female hormone estrogen, and so may be demasculinizing men and increasing the risk of breast and uterine cancer in women.

But that's just the beginning of the problem.

In 1960, routine feeding of antibiotics to farm animals was almost unknown. Antibiotic administration to meat animals has increased so much since then, however, that today over 55 percent of all antibiotics manufactured in the United States are put into animals or animal feed. This has turned our livestock into a vast breeding ground for antibiotic-resistant microorganisms.

The American pharmaceutical and meat industries don't consider this a problem (nor do the politicians to whom they contribute millions of dollars a year) and continue to support the routine administration of these drugs to dairy and meat animals. But that stance is hardly supported by any "common knowledge" in science; the European Economic Community (Common Market) has banned American meat products grown with antibiotics.

Why? The research that so concerned the Europeans showed that in 1960 only 13 percent of American human staphylococcus infections were resistant to penicillin. By 1988, however, the number of penicillin-resistant staph infections in Americans had exploded to over 90 percent. (Muppets creator Jim Henson was killed by such a drug-resistant infection, for example, despite his substantial wealth and his access to the most sophisticated and expensive health-care system in the world.)

And it's not just on land. Ocean studies specialist James W. Porter of the University of Georgia points to the explosion of human virus and bacteria filling the waters of the planet's oceans, killing off coral reefs and spreading disease among humans. He projects that 20 to 30 percent of coral reefs are at risk, with a 446 percent increase in infections since 1996 among the reefs he has been monitoring off the Florida coast. University of South Florida researcher Joan B. Rose points out that between 20 and 24 percent of all people swimming in Florida's coastal beaches become infected with viruses that can cause heart disease, ear infections, sore throats and eyes, meningitis, gastrointestinal disease, hepatitis, and diabetes. About 1 percent, she says, become chronically infected. Similarly, a sampling of shellfish from New York waters found over 40 percent infected with human pathogens, and a sampling of water from the Waikiki Beach in Hawaii found more than a third of the samples tested were infected with human viruses.

Vermont ice cream manufacturer Ben & Jerry's once sued the government. They wanted to put on the label of their ice-cream packages that the milk they use is from cows free of synthetic growth hormones or unnecessary antibiotics. But the government considers this information something that is so irrelevant to consumers that— at the well-financed suggestion of lobbyists for the drug manufacturers—they've passed laws banning dairy product packagers from mentioning whether *or not* these hormones are fed to their animals. And when news reporters wanted to report on the hormone issue, they were told to either lie to the public or be fired.

At least now there's enough rainfall and a reasonable enough climate in Vermont to raise the cows to produce high-quality milk for Ben & Jerry's ice cream. The early warning signs climate scientists see worldwide indicates that the "good weather" of the past few thousand years may be about to change, again in response to human activities.

Climate Changes

 One of the extraordinary things about human events is that the unthinkable becomes thinkable.

—SALMAN RUSHDIE (b.1948)

One recent July afternoon we had an electrical storm here in central Vermont that was so severe it took out two of my computers and blew circuit breakers throughout the house. Our home wasn't unique: many families lost most or all of their electrical appliances.

Larry, a fellow we'd hired to do some repair work on our half-mile-long driveway, stood atop a hill with me a week after the storm and told how his wife had been thrown across the room from an electric shock she received touching their screen door during the storm. "It's not normal weather here," he said. "Used to be that Vermont weather was famous for always changing, always unpredictable, but the last few years have been like nothing before."

The insurance industry agrees with Larry.

The decade of 1980–1989 was the costliest in history for insurance claims caused by "acts of God," with total claims of over $50 billion. But just the first five years of the 1990s saw claims of over $162 billion, prompting the insurance industry to issue an unprecedented call for a decrease in carbon dioxide emissions from industry.

On July 11, 1996, the Associated Press ran a story worldwide reporting that the growing season of the Northern Hemisphere has lengthened by about a week over the period from 1976 to 1996. Researcher Charles Keeling of the Scripps Institution of Oceanography in La Jolla, California, was quoted in the study, based on an article in the journal *Nature,* as saying that this was probably the result of global warming.

In the past decade, the science of phenology—the study of organisms' response to changes in seasons and climate—has grown both in interest and information. The data that phenologists have gathered, particularly since the mid-1990s, has been alarming. One of the big problems has come about because some species time their "waking up" in the spring and their outbound migrations or hibernations in the winter by temperature trends, while others time theirs by seasonal changes in the number of hours and minutes of sunlight. When species that rely on these two different systems are interdependent, and the temperature cycle changes but the sunlight cycle stays the same, the result can be disaster.

Consider the "great tit" (*Parus major*), a European bird about the size of a North American chickadee that doesn't migrate during the winter. An article in the February 22, 2003, *Proceedings of the Royal Society of London B,* recounted in the March 8, 2003, *Science News,* documents how the birds require an ample supply of the caterpillars of the winter moth (*Operophtera brumata*) to feed their young when they hatch in the spring. The moth's caterpillars, in turn, need fresh young buds of the European oak tree *Quercus robur.* The caterpillars' hatching is more sensitive to temperature changes than the oak tree, resulting in the caterpillars hatching "2 to 3 weeks before the oak buds open." This, Sid Perkins of *Science News* notes, is "not good for the caterpillars, which typically can survive only 2 or 3 days—and absolutely no more than 10 days—without food." A study of the great tits and the winter moths in the Netherlands found these nonsynchronized changes in the hatching time of the caterpillars; the budding of their food source, the oak trees; and the hatching of the baby birds was setting up a potential disaster for both the moths and the birds.

Similar nonsynchronous problems are noted between the Cassin's auklet seabird that lives along the North American coast between northern Mexico and southern Alaska and its primary food source, a tiny orange crustacean known as *Neocalanus cristatus.* The *Neocalanus* has, as a result of climate-change-induced variations in Pacific water temperatures, altered the very short (two-month) window of time when it lives near the ocean's surface. The birds, however, haven't changed their egg-laying cycles, which are apparently more timed to the sun than the temperature. The result was that the birds' food source "had come and gone by the time the birds hatched. As a result, auklet parents returned to their burrow nests with gullets filled with larval rockfish—'an unappetizing gray mush,' Bertram notes—instead of *N. cristatus,* the preferred prey. Accordingly, large numbers of auklet chicks died that year, and those that survived grew more slowly than normal."

Altogether, the study summarized 61 long-term research projects that analyzed almost 700 species over the past 50 years. "Those research projects show that some animals have been reaching life cycle milestones, such as breeding and egg laying, an average of about 5 days earlier per decade. The budding and blooming of trees, however, had advanced only 3 days per decade." The summary from a report in the January 2, 2003, *Nature* cited in *Science News* was that this strongly supports the idea "that climate change is already affecting ecosystems worldwide."

A report by the Rocky Mountain Biological Laboratory in Crested Butte, Colorado, found that yellow-bellied marmots (also known as groundhogs or woodchucks) are ending their hibernations 38 days earlier than they had 23 years ago. *Science News* reports "Global warming may be cutting short the marmots' long winter naps, says David W. Inouye, a biologist at the University of Maryland, College Park."[1]

Global warming is one of those things that everybody seems to have an opinion about but few people understand. The Earth's atmosphere is made up of gases and water vapor, and the primary gases are nitrogen (78 percent) and oxygen (21 percent). Argon is the next most common gas, yet it along with all other gases account for only about 1 percent of the total atmosphere, so you can see that the infamous carbon dioxide is only present in very small quantities in the atmosphere.

Oxygen and nitrogen allow light and heat to pass through them rather easily. Carbon dioxide, however (which is a small fraction of that remaining 1 percent of the dry atmosphere), behaves quite differently. It acts like a blanket or quilt around the Earth, trapping heat in and below the atmosphere. Gases that behave like this are often referred to as "greenhouse gases" because they act like the glass on a greenhouse, trapping the sun's heat and keeping the plants inside warm. (Methane, which also contains carbon, is another greenhouse gas.)

For example, while the planet Venus is only 27 percent closer to the Sun than the Earth, its surface temperature is over 700° Fahrenheit. Given Venus's proximity to the Sun, its surface should be substantially cooler than that, but it has an atmosphere that is rich in carbon dioxide: a greenhouse gas. Therefore, the temperature of the planet's surface is held 700 degrees hotter than it would be if its atmosphere was made up of the same 99 percent nitrogen and oxygen as our planet's atmosphere.

One of the primary roles that carbon dioxide plays in our atmosphere is to regulate the temperature of the planet's surface. If there were substantially less CO_2, the Earth's surface would become covered with ice. If there were more than there is today, the surface would warm (as it has been steadily doing since around 1890, because of the rapid increase in carbon released into the atmosphere by our burning of fossil fuels).

At previous times in the Earth's history, there was much more carbon dioxide in the atmosphere than there is today. During the aptly named Carboniferous Period, over 300 million years ago, the planet was warm to the point of hot nearly worldwide, and plant life flourished in the warm and carbon dioxide–rich environment.

The combination of heat and carbon dioxide led to such an explosion of plant life that huge amounts of carbon were extracted from the atmosphere and converted into vegetation. This, in turn, led to a decrease in the levels of carbon dioxide in the air, which caused a gradual cooling of the planet because the carbon dioxide "blanket" was thinned out.

The two primary ways that carbon is removed from the atmosphere are through the growth of trees and coral reefs. These two "carbon sinks" act as a vast reservoir for carbon, keeping it out of the atmosphere. While coral is more permanent, forests will nonetheless hold carbon for centuries. And when forests become fossilized and are converted to oil or coal, they can hold the formerly atmospheric carbon for millions of years.

It took hundreds of millions of years for trees—ranging from ancient to modern forests and plants—to pull billions of tons of carbon out of the Earth's atmosphere and store it in the Earth.[2] The resulting decrease in greenhouse carbon dioxide levels in the air along with other factors produced the climate we enjoy now, one very different from the climates in the past. Modern forests account for the most massive of the current atmosphere carbon storage systems. Scientists point out (in the *Nature* article referenced earlier) that there's a measurable annual fluctuation in atmospheric levels of carbon dioxide that has to do with plants growing during the summer (and pulling down carbon from the air) and shedding their leaves in the fall and winter (and thus releasing carbon back into the air as the leaves decompose or are burned). The swings in this cycle have increased as much as 40 percent over the past 30 years, according to scientist Keeling, probably because of the weeklong extension of the growing season in the northern hemisphere.

And the rate of "stored" carbon release is accelerating at an incredible rate. During just a 10-year period—the decade of the 1980s—scientists estimate that fully 15 percent of the new carbon dioxide in the atmosphere was released as the result of one single human event: the burning of the tropical rainforests in the Americas, mostly to make ranchland for cattle.

This has caused some dispute in the scientific community about the impact of deforestation, because the rate of tree loss hasn't perfectly matched the rate of increase in atmospheric carbon dioxide. The carbon dioxide hasn't increased as fast as predicted due to its release into the atmosphere by the burning of the rainforest trees, which has caused some skeptics of the global-warming theory to ridicule the idea that deforestation may lead to increases in greenhouse gases. They point out that fully a quarter of the carbon dioxide that is emitted by the burning of trees appears to have vanished from the atmosphere, calling into question the original calculations of release of CO_2 or the mechanism by which it's stabilized.

But research conducted by scientist Jeffrey Andrews of Duke University and reported by him at the 1996 meeting of the Ecological Society of America explains this, and shows that trees are even more critical to maintaining steady levels of atmospheric carbon than anybody had previously thought.

Andrews examined water in the ground around trees, and at a distance from trees, and found that the water near trees contains higher levels of carbon dioxide. The trees, it seems, draw large amounts of carbon dioxide from the atmosphere and pump it down into the soil. From here it leaches into groundwater, which keeps it from quickly escaping back up into the atmosphere. In some cases, groundwater percolates downward and is trapped in the earth for tens of thousands of years, holding its store of trapped carbon dioxide. (Such water, when liberated centuries later, is "naturally carbonated"; this process is created by trees.)

To demonstrate his observation, Andrews sprayed trees in a forest reserve in North Carolina with carbon dioxide, increasing their leaves' exposure to the gas by 50 percent over what they'd normally experience. Then he tested soil from a level about three feet deep below these trees. The CO_2 concentrations had risen by 25 percent.

Andrews said that living trees, catching surplus carbon dioxide from the burning of cut trees and fossil fuels, were trapping in the soil and storing as much as 20 percent of the missing carbon dioxide, and the carbon dioxide trapped in groundwater may remain stable for thousands of years.

While on the surface this seems like good news, meaning that the atmosphere isn't so rapidly affected by deforestation, the long-term implications are ominous. So long as there is a certain (currently unknown) percentage of trees alive, they'll be able to sink-out the excess carbon dioxide by putting that carbon dioxide into groundwater.

But when the loss of forests drops to the point where the remaining living trees cannot absorb the extra carbon dioxide, the result could be a crashing domino effect with a very sudden increase in atmospheric CO_2. Levels would increase slowly but steadily until that last few acres are cut, and then there would suddenly be an unprecedented increase, leading to a profound change in global climate, perhaps over a period as short as just a few years.

The United Nations convened a congress of 2,500 of the world's leading scientists in the area of meteorology, ecology, geology, and other Earth sciences who had been researching these issues for years. The U.N.'s Intergovernmental Panel on Climate Change (IPCC) concluded that we are, indeed, facing a crisis that may well be of biblical proportions as a result of global warming produced by increased greenhouse gases in the atmosphere.

When we look up at the sky, it's easy to think that it's infinite, that it goes on forever, and that it would be nearly impossible to damage that vast vault of blue. Yet, as Bill McKibben points out so articulately in *The End of Nature,* the distance between the ground (at sea level) and the upper edge of the troposphere, the part of our atmosphere that supports almost all life on Earth, is only about six miles.[3] That's all that we have above and around us, just those narrow six miles of air, and crowded in and below that is every form of terrestrial life.

And, as a consequence of global warming, even that thin protective layer above us is dropping. "The sky is actually falling," noted a September 26, 1998, *Science News* article reporting on measurements taken of the height of the upper atmosphere between 1958 and 1998. Bouncing radar beams off the ionosphere—the layer of charged air particles in the uppermost atmosphere—scientists discovered that "the average ionosphere height dropped by 8 kilometers," an effect predicted by climate change models. As carbon dioxide levels, principally from burning fossil fuels, increase in the lower atmosphere, they both trap heat into the Earth and prevent that heat from warming the upper atmosphere. The result, according to computer models that had predicted this effect before it was measured, would be a cooling of the uppermost atmosphere by as much as 50° Celsius. Cool air settles and thickens, thus lowering the top of the sky.

Two hundred years ago, the thin layer of air above us contained an average of 280 parts per million of carbon dioxide.

Looking at the rings of 120-year-old trees in the Vermont mountains, forestry experts found regular patterns of growth every year for the first 30 years or so. Then the oil- and coal-fired factories of the industrial Midwest, the Tennessee and Ohio river valleys, and the construction lines of Detroit came on-line in the 1920s. And the rings began to change.

First they found that the trees grew faster, as a result of the extra carbon dioxide, which is a food for the trees. But that fast growth increased the rate at which the trees "exhaled" water vapor, and increased their need for rainfall . . . which didn't come along with the higher CO_2 levels.

In addition, the rainfall became acidic, which changed the nature of the mineral balance in the soil, leaching out the alkaline minerals like calcium, and liberating highly toxic aluminum. The result was the killing off of root structures by toxic metals and the weakening of the trees from lack of calcium and other alkaline minerals.

And there were toxic metals in those clouds from the factories and power plants. Substances like vanadium, zinc, mercury, lead, and other toxic or heavy metals—previously totally absent from the trees' earlier growth—began to show up in the rings grown in the years following the industrialization of America. It rose gradually from the turn of the century until the 1950s, and then the rate of accumulation of these toxins exploded.

So the trees began to die. According to research done by University of Vermont's Dr. Hubert Vogelmann and reported by Charles E. Little in his brilliant book *The Dying of the Trees,* the rate of tree death in the region of Vermont he studied had so accelerated that just in the 14 years from 1965 to 1979 over 40 percent of the red spruce have died, 73 percent of the mountain maples, 49 percent of the striped maples, and 35 percent of the sugar maples, the tree that most people visualize when they think of Vermont.[4]

Each year, because of our consumption of oil, gas, and coal, we're pumping more than six billion tons of heat-trapping carbon dioxide into that thin layer of atmosphere—so much that in just the past 20 years the concentration of CO_2 in the atmosphere has increased from 280 parts per million to over 370.9 parts per million, the highest

Figure 3. Atmospheric concentration of CO_2

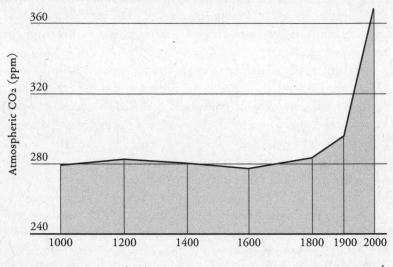

Source: American Institute for Physics http://www.aip.org/pt/vol-55/iss-8/captions/p30cap2.html[5]

level in 420,000 years. Within a few more decades, it's projected to exceed 500 parts per million, thus dramatically warming the planet.

But how warm? According to the scientists of the U.N.'s IPCC, at least 3° to 4° Celsius (5° to 7° Fahrenheit), and possibly as much as 7° Celsius (10° Fahrenheit).

"What's so bad about that?" many people ask. "Three degrees is nothing, and if that warms up the climate of Michigan or Maine, wouldn't that be better for the growing season, recreation, and everything else?"

Unfortunately, it's not that simple. As climatologist Ken Caldeira told *Discover* magazine in April 2003, just a 2° change over a hundred-year period—a conservative estimate of what may happen—is the same as moving worldwide climate bands over 250 miles toward the poles, the equivalent of 30 feet each day. "Squirrels might be able to move at those kinds of rates," Caldeira said, "but an oak tree can't."

Figure 4. Global average temperature

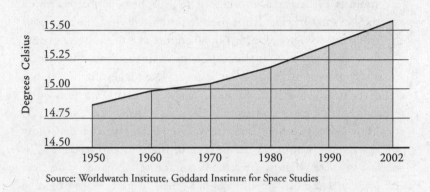

Source: Worldwatch Institute, Goddard Institute for Space Studies

Climate change driven by increasing carbon dioxide levels already appears to be producing huge swings in weather all over the planet, because heat is energy, and increased heat in the atmosphere means increased energy in the atmosphere. This increased energy makes for less stable and more violent weather worldwide.

The April 26, 2003, issue of Britain's *New Scientist* magazine arrived at my home the same week in May that the worst tornadoes in U.S. history were taking apart big chunks of the American Midwest. The article titled "Here Comes the Rain" opened with, "As the world gets warmer, it is getting wetter. And one of the main conclusions reached at Europe's largest ever earth science conference was that we are less prepared than ever." The article points out that as the overall temperature of the atmosphere increases, so, too, does its ability to hold ever-larger quantities of water. This increase in moisture and air density drives ever-more-powerful storms, and when the water is dropped from the skies, it makes for record-breaking floods.

When I was in Australia in 2002, as wildfires were ravaging New South Wales and licking up against Sydney, the evening news reported in nearly every newscast how the fires were caused by drought that climate scientists had previously predicted would be the unstoppable result of global warming. However, not a single news report in the U.S. on radio, TV, newspapers, or the Internet mentioned how the flooding in the midwestern U.S. had been predicted just months earlier at a conference of European global warming experts. Neoconservative front men for the fossil fuel industry have effectively cowed news reporters in all sectors of American reporting.

And while the U.S. is sticking its head in the sand, the climate scientists reported on in the *New Scientist* were suggesting that things could get much, much worse in a far shorter time than previously realized. This is because methane is 10 times more effective than CO_2 at causing global warming, and the warmer, wetter weather caused by global warming stimulates the growth of marshes and wetlands, "as bacteria in these swamps are a major source of methane." The clear and present danger is the same as happened in past cycles of global cooling and warming: "If an initial nudge in climate made the world wetter, that could have extended wetlands and triggered further, rapid warming."[6]

The projected increase of Earth's mean temperatures by 3° to 4° Celsius (5° to 7° Fahrenheit) is startlingly like the planetary change between the last ice age and now. (The Ice Age ended and the oceans rose 500 feet because the planet warmed by 7° Celsius 10,000 years ago.)

While industry front groups like the official-sounding National Center for Public Policy Research delight in issuing neoconservative propaganda like their April 2001 report titled "Carbon Dioxide Is Good for the Environment" (widely circulated among and by Republican legislators), real scientists know it's a lie. For example, a December 5, 2002, study published by Stanford University ("High Carbon Dioxide Levels Can Retard Plant Growth" by Mark Shwartz) found that "elevated atmospheric carbon dioxide actually reduces plant growth when combined with other likely consequences of climate change—namely, higher temperatures, increased precipitation or increased nitrogen deposits in the soil." Yes, plants breathe in carbon dioxide as if it were food, but the simplistic view promoted by industry and their captive politicians that "more food is good" is simply wrong, even though it plays well on venues like right-wing talk radio.

* * *

The oil industry denies this, but insects and wildlife can't be fooled: they're moving, because the weather is changing.

When we moved to Atlanta in 1983, we rarely saw mosquitoes. When they did come out, they were relatively large and clumsy, flew slowly, and were a pest only in the early evening. Sometime around 1990, however, things changed. Mosquitoes began to harass us, smaller and faster ones that fed during the daytime and in broad daylight. They were out for much of the year, and displayed a tenacity I'd never seen before in an American mosquito.

What I'd noticed, it turned out, was the northbound migration of the *Aedes aegypti* mosquito, a normal tenant of tropical climates that first came to Florida in the late 1950s. According to investigative journalist William Blum, author of the 2000 book *Rogue State,* between 1956 and 1958 "The Army, wishing to test 'the practicality of employing *Aedes aegypti* mosquitoes to carry a BW [biological warfare] agent,' released over wide areas hundreds of thousands, if not millions, of this mosquito, which can be a carrier of yellow fever and dengue fever." The insects were released, according to Blum, in Savannah, Georgia, and Avon Park, Florida. An entomologist at one of the universities in Atlanta told me, "It's a day-feeder, fast breeder, fast flyer, and smaller and smarter than our natural local mosquitoes. And, unlike them, it will carry yellow fever, dengue hemorrhagic fever, Japanese encephalitis, and malaria."

The next spring, we read reports in the paper about the first recent case of malaria in South Carolina, and as the climate of North America warms, we'll see it progress northward as it is doing today in other parts of the world.

Similarly, according to 1995 research done in the Netherlands and Great Britain, the global warming predicted by the ICCP, would lead to a doubling of malaria-carrying mosquito populations in tropical regions. But the danger is much greater for the temperate climates, the researchers point out: the United States, Europe, Russia, and China face a *hundredfold* increase in risk. The additional 50 to 80 million cases of malaria would be particularly devastating in those regions, such as the United States, where the population has not been exposed to the disease for generations and thus has not developed any immunity to the disease at all.

Dengue fever is carried by the same mosquitoes. Often called breakbone fever because it causes such extreme pain in the bones and joints, and severe headaches, it has recently moved as far north as Puerto Rico, where 15,000 people were infected in a recent epidemic. While dengue is debilitating and one of the world's most painful non-lethal diseases, it has recently mutated into a far more deadly form called dengue hemorrhagic fever. DHF begins as red spots and the fever and pain characteristic of dengue, but then begins to break down the internal capillary system of the body, so that massive internal bleeding of the brain, lungs, and intestines takes place. Blood pours from the nose and rectum, and the patient dies from internal bleeding. Between 1981 and 1985 there were an average of 100,000 cases of HDF per year, but this number more than quadrupled during the "hot years" of 1986 to 1990 to over 450,000 cases a year.

Another disease carried by mosquitoes that has exploded recently, particularly in the region of New Jersey, Massachusetts, and New York, is eastern equine encephalitis. Formerly a rare disease in humans, this virus kills about 60 percent of people who contract it through the bite of a mosquito, and recent outbreaks have led to aerial spraying of insecticides in mid-Atlantic states. At the same time, the *Aedes aegypti* mosquito is helping fuel a growing epidemic of West Nile virus across the eastern and central United States.

On the other coast, seabirds from California to Oregon have been devastated by a 2° Fahrenheit average water temperature increase since 1950, which has redirected the flow of cold, nutrient-rich ocean waters away from the coast. That, in turn, has led to a 40 percent decline in the zooplankton, which is the food supply for shrimp and other small marine creatures. Fish and squid live on the shrimp, and the birds live on the fish. In just the years between 1987 and 1994, for example, four million sooty shearwater birds died along that coast, reducing their population by over 90 percent.

At Glacier National Park, Montana, the 30,000-year-old glacier is melting so rapidly that scientists predict it'll be gone within 30 years. In northern Michigan, the migrating red-winged blackbirds, Canada geese, broad-wing hawks, and hummingbirds are arriving three weeks earlier than they were in 1965. The Gangotri glacier is retreating 30 meters per year in India, the Qori Kalis glacier is retreating 100 feet per year in Peru, and Russia's Caucasus Mountains have, over just the past 100 years, lost fully half of all the glacial ice.

A report in the August 30, 1997, *Science News* quoted researchers in the United Kingdom, who had analyzed over 74,258 records of 65 bird species, and found that all except one species were laying their eggs "nine days sooner than in 1971," showing clear evidence of the effects of global warming. The researchers expressed concern that this may throw off the normal feeding and life cycle of the birds, although it will be some time before the full effects are known.

At the same time, severe weather is cranking up, with both drought-driven wildfires and warmer-atmosphere-driven tornadoes, hurricanes, and tropical storms amping up in both frequency and ferocity around the world.

Despite the attempts of the American oil and coal industry to obfuscate the scientific facts (as chronicled in the book *The Heat Is On* by Ross Gelbspan), people are waking up.

The Garden of Eden and the flood

This isn't the first time, of course, that changes in the planet's climate have threatened humans. We can find what may be two stories of this in the Bible.

The Garden of Eden, according to Genesis 2:10–14, was located at the confluence of four rivers: the Pison, Euphrates, Gihon, and Hiddekel. The Pison is believed by many archaeologists to be the present-day Wadi Batin, a dry riverbed. The Hiddekel is the archaic name for what we today call the Tigris, and the Euphrates is still called the Euphrates. The Gihon is believed to be the present-day Karun River, which, before it was dammed, flowed from the headlands of Iran to form the delta of the present-day Persian Gulf.

If you track these four rivers back, you find that the point where they meet is several miles out into the present-day Persian Gulf. While this area is today underwater, it was a rich and fertile forestland around 10,000 years ago, just before the last ice age ended. At that time, so much water had been taken from the oceans and locked up in glacial ice that the world's oceans were 500 feet lower than they are today. At that time, most humans alive on the Earth were, as far as we can tell, hunters and gatherers, living the "leisure lifestyle" of the !Kung/San, Hottentots, Shoshone, and others whom we'll meet later in this book.

Then the climate changed and the people living in that area were forced to move away from the rising ocean waters. In their new lands, they changed their lifestyle from hunting/gathering to herding and agriculture, the respective professions of Adam and Eve's first two sons. In this, the writers of Genesis say, humans were thus expelled from Eden and "cursed" to work the fields, their faces "covered with sweat."

Others weren't able to simply move on to avoid the rising waters, however. The *Epic of Gilgamesh,* written several thousand years before the Bible, has in it the story of Utnapishtim, a righteous man who was warned in a dream to build an ark. He did so, gathering together his family and two each of every animal, just as the rains began. His ark floated him, his family, and all his animals to safety, then it finally came to rest on a mountaintop. As Colin Tudge points out in *The Time Before History,* "The parallels [of the story in the *Epic of Gilgamesh*] with the account of Noah in Genesis are exact."[7]

The Bible points out, in Genesis 11:31, that Abraham gathered together his family and started his travels to what would later become known as Israel from the area of Ur, which, according to Tudge, is near where Gilgamesh had reigned in Uruk many years earlier. Most likely this is the source of the story of the biblical flood, or they may be parallel stories. In either case, the stories of the arks in the *Epic of Gilgamesh* and the Bible both may well point to a historical occurrence that took place at the end of the last ice age. The worldwide temperature rose a full 7° Celsius, causing so much ice to melt that the oceans rose 500 feet and the atmosphere was probably saturated for some time with moisture, producing awesome monsoon-like rains, probably lasting years. (Tudge cites evidence that this radical climatic shift took place in as little as 20 years. Ice-core samples drilled in Greenland in 1999 indicate the shifts may have taken as little as two years.) Boat owners would have been at a significant survival advantage, and at least one or two may have lived to tell their tale.

* * *

The second day of January 1999 saw a milestone in the efforts to protect the Brazilian rainforest, an area half the size of the United States, which contains two-thirds of the planet's non-glaciated fresh water. On that date, it was reported that the Brazilian government, bowing to intense pressures from the International Monetary Fund (IMF) to cut spending, slashed their budget for protecting the rainforest.

In Brazil's rainforests are hundreds of indigenous tribes, and the $250 million rainforest-protection program's first priority was to survey a 25-million-acre area that would be kept, in theory, forever intact for their use alone. By cutting the $250 million budget down to $6 million, the program is barely kept alive, and now nothing substantive will be done to protect the trees or the people. In the meantime, the forest is being overrun by an army of loggers, ranchers, miners, farmers, and evangelists bent on the "salvation" of the "heathen" tribes, and the trees are being cut and burned at a rate of more than 200,000 acres a day.

This forest, with 20 percent of the world's fresh water, is one of the planet's most important sources of atmospheric water vapor, second only to the oceans, and therefore has tremendous impact on the planet's weather patterns. It's also one of the planet's most important carbon sinks, keeping carbon stable in the trees. Now, as vast areas that were once rainforest become tree-denuded strip mines and grazing land for cattle, carbon is released into the air instead of water vapor, contributing to global warming and producing changes in the weather patterns of Europe, the Middle East, and northern Africa.

South of Brazil, 70 percent of the world's fresh water is locked up in Antarctica, where it has been held as ice for hundreds of thousands of years. Covered by ice sheets that are often more than 3 miles high, the continent of Antarctica covers 5.4 million square miles, larger than the size of China and India combined. If the ice of Antarctica was to melt and slide off that continental landmass and into the sea, oceans would rise significantly around the world.

This appears to be happening.

In April 1999, it was reported that researchers with the British Atlantic Survey in Cambridge and the University of Colorado analyzed data on the ground and from satellite photos. They found conclusive evidence that global warming has lengthened the annual melting season of Antarctica by three weeks, producing drastic changes to that continent's ice shelves. The Wilkins and Larsen B ice shelves, for example, are—to quote the scientists—in "full retreat." In just the four months from November 1998 to February 1999, over 420,000 acres of the Larsen B shelf's total 1.7 million acres caved away. In just the month of March 1999, the Wilkins shelf lost over a quarter million of its three million acres, and the process has dramatically speeded up in the years since.

British Antarctic Survey researcher David Vaughan was quoted in the Associated Press as saying that within just a few short years "much of the Wilkins shelf will be gone."

The reason? Average temperatures in Antarctica—relatively stable since the dawn of humanity—have risen 4.5° Fahrenheit since 1950, pushing the summer temperatures there above the critical freezing point of 32° Fahrenheit.

A new ice age?

In the context of all this discussion about global warming, it may seem odd to bring the conversation around to a new ice age. But that may well be the greatest threat the world is facing today, particularly North America and northern Europe.

If you look at a globe, you'll see that the latitude of much of Europe and Scandinavia is the same as that of Alaska and permafrost-locked parts of northern Canada and central Siberia. Yet Europe has a climate more similar to that of the United States than northern Canada or Siberia. Why?

It turns out that our warmth is the result of ocean currents that bring warm surface water up from the equator into northern regions that would otherwise be so cold that even in summer they'd be covered with ice. The current of greatest concern is often referred to as the Great Conveyor Belt.

Although most ocean currents are driven by wind and the Coriolis effect (named after French mathematician Gustave Gaspard Coriolis, 1792–1843). The Coriolis effect is the result of the spin of the Earth and accounts for why riverbeds are deeper on one side and north-south railroad tracks always wear out faster on one side than the other. The spin of the Earth produces a constant but small force that drives our winds and most ocean currents, particularly in the Pacific Ocean and in the continuous oceanic circle around Antarctica.

The Great Conveyor Belt, however, while shaped by the Coriolis effect, is mostly driven by the greater force created by differences in water temperatures and salinity. The North Atlantic Ocean is saltier and colder than the Pacific, the result of it being so much smaller and locked into place by the Northern and Southern American Hemispheres on the west and Europe and Africa on the east. Winds hit it after passing thousands of miles of land, and, unlike the Pacific, it's not large enough to have widely circulating currents to distribute equatorial heat. And, because the arctic ice cap is frozen solid, there's no way for the Coriolis effect to push Atlantic water up and over the top of Europe and Siberia. (In the south, the ocean all the way around Antarctica is open, so the southern Atlantic doesn't have this problem.)

As a result, water evaporates out of the North Atlantic, leaving behind salt, and the cold continental winds off the northern parts of North America cool the waters. Salty, cool waters settle to the bottom of the sea, most at a point a few hundred kilometers south of the southern tip of Greenland, producing a whirlpool of falling water that's 5 to 10 miles across. While the whirlpool rarely breaks the surface, during certain times of year it does produce an indentation and current in the ocean that can tilt ships and be seen from space.

The Atlantic's undersea river

This falling column of cold, salt-laden water pours itself to the bottom of the Atlantic, where it forms an undersea river 40 times larger than all the rivers on land combined, flowing inexorably south down to and around the southern tip of Africa, where it finally reaches the Pacific. Amazingly, the water is so deep and so dense (because of its cold and salinity) that it often doesn't surface in the Pacific for as much as a thousand years after it first sank in the North Atlantic off the coast of Greenland.

The outflowing undersea river of cold, salty water makes the level of the Atlantic slightly lower than that of the Pacific, drawing in a strong surface current of warm, fresher water from the Pacific to replace the outflow of the undersea river. This warmer, fresher water slides up through the South Atlantic, loops around North America, where it's known as the Gulf Stream, and ends up off the coast of Europe. By the time it arrives near Greenland, it's cooled off and evaporated enough water to become cold and salty and sink to the ocean floor, providing a continuous feed for that deep-sea river flowing to the Pacific.

These two flows—warm, fresher water in from the Pacific, which then grows salty and cools and sinks to form an exiting deep sea river—are known as the Great Conveyor Belt.

Amazingly, the Great Conveyor Belt is the only thing between comfortable summers and a permanent ice age for Europe and the eastern coast of North America.

Ancient tropical times

It wasn't always this way. Up until around three to four million years ago, the Isthmus of Panama was deep underwater, and warm, fresh waters from the Pacific flowed into the Atlantic at the equator. These warm waters then flowed up around Europe and over the top of Siberia across the then-open Arctic Ocean. The result was that both North America and Europe had millions of years of very, very mild weather. The winters just got down into the 50s Fahrenheit, and the summers just got up into the 80s. The total calories of heat from the sun hitting the Northern Hemisphere were the same as today, but because of this moderating flow of water, they were evenly distributed across the world, producing warm winters and gentle summers. (It was this way all the way back to the era of the dinosaurs and before.)

Then, starting around 4.6 million years ago, things began to change. The thin landmass that connects North and South America, deep underground for those millions of years, began to rise, finally concluding in its current form several hundred feet above sea level around 1.9 million years ago. When the Isthmus of Panama rose up from the sea bottom and formed a continuous landmass between North and South America, it blocked the flow of warm, fresh water from the Pacific. When it blocked that water, it also blocked the main source of moderating heat for northern Europe and the northeastern part of North America. The main blockage seems to have occurred around three million years ago, as that's generally regarded as the beginning of the Great Ice Age.

And, although most people who aren't climatologists don't realize it, we're still in that ice age.

Much of the science that I'm sharing with you here was unknown as recently as 20 years ago. Then an international group of scientists went to Greenland and used newly developed drilling and sensing equipment to drill into some of the world's most ancient accessible glaciers. Their instruments were so sensitive that when they analyzed the ice core samples they brought up, they were able to look at individual years of snow. The results were shocking.

Prior to the last decades, it was thought that the periods between glaciations and warmer times in North America, Europe, and North Asia were gradual. We knew from the fossil record that the Great Ice Age period began a few million years ago, and during those years there were times where for hundreds or thousands of years North America, Europe, and Siberia were covered with thick sheets of ice year-round. In between these icy times, there were periods when the glaciers thawed, bare land was exposed, forests grew, and land animals (including early humans) moved into these northern regions.

Most scientists figured the transition time from icy to warm was gradual, lasting dozens to hundreds of years, and nobody was sure exactly what had caused it. (Variations in solar radiation were suspected, as were volcanic activity, along with early theories about the Great Conveyor Belt, which, until recently, was a poorly understood phenomenon.)

Looking at the ice cores, however, scientists were shocked to discover that the transitions from ice age–like weather to contemporary-type weather usually took only two or three years. Something was flipping the weather of the planet back and forth with a rapidity that was startling.

It turns out that the ice age versus temperate weather patterns weren't part of a smooth and linear process, like a dimmer slider for an overhead lightbulb. They are part of a delicately balanced teeter-totter, which can exist in one state or the other, but transits through the middle stage almost overnight. They more resemble a light switch, which is off as you gradually and slowly lift it, until it hits a midpoint threshold or "breakover point," where suddenly the state is flipped from off to on and the light comes on.

It appears that small (less than .1 percent) variations in solar energy happen in roughly 1,500-year cycles. This cycle, for example, is what brought us the "Little Ice Age" that started around the year 1400 and dramatically cooled North America and Europe (we're now in the warming phase, recovering from that). While the ice in the Arctic Ocean is frozen solid and locked up, and the glaciers on Greenland are relatively stable, this variation warms and cools the Earth in a very small way, but doesn't affect the operation of the Great Conveyor Belt that brings moderating warm water into the North Atlantic.

In millennia past, however, before the Arctic totally froze and locked up, and before some critical threshold amount of fresh water was locked up on Greenland and in other glaciers, these 1,500-year variations in solar energy didn't just slightly warm up or cool down the weather for the landmasses bracketing the North Atlantic. They flipped on and off periods of total glaciation and periods of temperate weather.

And these changes came suddenly.

For early humans living in Europe 30,000 years ago—when the cave paintings in France were produced—the weather would be pretty much like it is today for well over a thousand years, giving people a chance to build culture to the point where they could produce art and reach across large territories.

And then a particularly hard winter would hit.

The spring would come late, and summer would never seem to really arrive, with the winter snows appearing as early as September. The next winter would be brutally cold, and the next spring didn't happen at all, with above-freezing temperatures only being reached for a few days during August and the snow never completely melting. After that, the summer never returned: for 1,500 years the snow simply accumulated and accumulated, deeper and deeper, as the continent came to be covered with glaciers, and humans either fled or died out. (Neanderthals, who dominated Europe until the end of these cycles, appear to have been better adapted to cold weather than *Homo sapiens*.)

What brought on this sudden "disappearance of summer" period was that the warm-water currents of the Great Conveyor Belt had shut down. Once the Gulf Stream was no longer flowing, it took only a year or three for the last of the residual heat held in the North Atlantic Ocean to dissipate into the air over Europe, and then there was no more warmth to moderate the northern latitudes. When the summer stopped in the north, the rains stopped around the equator. At the same time Europe was plunged into an ice age, the Middle East and Africa were ravaged by drought and wind-driven firestorms.

Nobody is sure why the Great Conveyor Belt ocean current would shut itself down, sometimes over cycles of 1,500 years, sometimes over 18,000 and even 125,000 years. And nobody's sure why around 10,000 years ago the cycles stopped and the Great Conveyor Belt has been running continuously ever since. One widely accepted theory has to do with a combination of factors, starting with energy from the sun.

Over time, the solar-driven glaciation cycles built deeper and larger glaciers and icebergs, drawing more and more fresh water out of the oceans, leaving behind more and more salt. (Ocean water would evaporate, condense, and fall as snow, which was then held on landmasses as glaciers and as the Arctic ice cap.) Eventually, like somebody slowly lifting a dimmer light switch, a certain level of salinity was reached and a certain amount of the Arctic Ocean became blocked solid with accumulated ice. The balance between salinity, temperature, and range of ocean flow hit a critical threshold—the halfway point of our off-on light switch—and the Great Conveyor Belt was switched on in a way that was so far "on" that it would take a major change in conditions to turn it back off again. It's run continuously for the past ten thousand or so years.

Thus began what we call modern civilization.

Global warming and the end of civilization

If the Great Conveyor Belt, which includes the Gulf Stream, were to stop flowing, the result would be sudden and dramatic. Winter would set in for the eastern half of North America and all of Europe and Siberia, and never go away. Within three years, those regions would become uninhabitable and nearly two billion humans would starve, freeze to death, or have to relocate. Civilization as we know it probably couldn't withstand the impact of such a crushing blow.

And, incredibly, the Great Conveyor Belt has hesitated a few times in the past decade. As William H. Calvin points out in one of the best books available on this topic (*A Brain for All Seasons: Human Evolution and Abrupt Climate Change*):

> . . . the abrupt cooling in the last warm period shows that a flip can occur in situations much like the present one. What could possibly halt the salt-conveyor belt that brings tropical heat so much farther north and limits the formation of ice sheets? Oceanographers are busy studying present-day failures of annual flushing, which give some perspective on the catastrophic failures of the past. In the Labrador Sea, flushing failed during the 1970s, was strong again by 1990, and is now declining. In the Greenland Sea over the 1980s salt sinking declined by 80 percent. Obviously, local failures can occur without catastrophe—it's a question of how often and how widespread the failures are—but the present state of decline is not very reassuring.[8]

Most scientists involved in research on this topic agree that the culprit is global warming, melting the icebergs on Greenland and the Arctic icepack and thus flushing cold, fresh water down into the Greenland Sea from the north. When a critical threshold is reached, the climate will suddenly switch to an ice age that could last minimally 700 or so years, and maximally over 100,000 years.

And when might that threshold be reached? Nobody knows—the action of the Great Conveyor Belt in defining ice ages was discovered only in the last decade. Preliminary computer models and scientists willing to speculate suggest the switch could flip as early as next year, or it may be generations from now. What's almost certain is that if nothing is done about global warming, it will happen sooner rather than later.

Consider where we are at this point

- We're all made out of sunlight, and everything we depend on is fueled by sunlight.
- For hundreds of thousands of years we lived off of current local sunlight.
- Then we discovered ancient sunlight, buried in the ground, and began consuming it both for its heat and as a raw material to replace plant-made fabrics.
- "Capturing" the ancient sunlight increased our productivity, but it also increased our appetite for more. Worse, it enabled faster and faster population growth.
- It also led to climate changes, which are destabilizing populations of other life-forms.
- Now the last hours of ancient sunlight are within view, perhaps less than a lifetime away. And even if alternative energy sources are developed, they may actually worsen the problem (by adding more people) if our culture doesn't change along with them. So long as we use ancient sunlight, or any other resource, to conquer nature and convert natural areas into human habitations, we imperil ourselves as well as the rest of the world by our numbers, our wiping out resources, and our competition with all other species for space, water, and nutrients.
- If no immediate alternative to oil appears, and we can no longer produce (and deliver) food for all the extra people we have, what will happen? Let's look at how one huge country is addressing the question.

A Visit to a Country That's Planning
How to Survive: China

 In the arts of life man invents nothing: but in the arts of death he outdoes Nature herself, and produces by chemistry and machinery all the slaughter of plague, pestilence, and famine.
——GEORGE BERNARD SHAW (1856–1950)

The Chinese are planning to survive the future. I discovered, some years back, that they're far more intentional, and far more concerned, about it than are most other countries in the world. They're taking the long view, looking at food and energy and resources in decade-long and generation-long pictures, rather than the single-year or even single-quarter views that dominate American corporate decision-making and the politics of responding to last week's opinion poll.

In November 1986 I stood in Tiananmen Square wearing a tan raincoat with the collar turned up and a Cossack hat to protect my head from the freezing mist. For the previous week I'd been sharing a room with three doctors from Colombia and Japan at the world's largest acupuncture teaching hospital, spending my mornings memorizing Chinese words and acupuncture points and meridians, my afternoons sticking needles in patients out on the wards, and my evenings exploring the city. Today was my first day off in a week, and I was going to visit the places that were closed in the evenings.

The mist from the clouds feebly cleaned the air of the thick smell of millions of tiny coal fires: most residents of the city heated a single room with their weekly allotment of a few bricks of coal, and the smoke wove itself through the air, often making it painful to breathe. The cold from the damp air penetrated every muscle and bone of my body; for several days I'd felt perpetually chilled.

A gray, mottled layer of clouds pressed down on Beijing like a giant hand, hardly moving, just a few hundred feet above the Great Hall of the People. Tiananmen Square is a large, open expanse of concrete surrounded by buildings of importance. The Great Hall, Mao's tomb, the Museum of the People. Beside me stood Dr. Wu, a lean, tall, intense man in his late twenties who had offered to show me the city in exchange for an opportunity to practice his English.[1]

"Do you have children?" I asked him. We'd walked in silence for several minutes, as he'd thought we were being followed. Now there was nobody closer than 20 feet.

"Yes," he said, and sighed, as if recalling a distant and bittersweet memory. "My wife and I have a daughter. They live a day's journey from here."

"One child?"

He looked at the ground for a moment, scanning the concrete like a person afraid he might step on something fragile. "Yes. One child."

"What do you think of that policy?" I said.

He darted me a guilty look, and then glanced around us. "It is wise," he said. "It is necessary for the future of China."

"Do you want a son?"

He shrugged. "Everybody wants a son. The odds are fifty-fifty."

"Yet there are more boys than girls among the children." I'd seen the playground of the nursery school across the street from the hospital: about two-thirds of the children were boys.

He shook his head, a sharp jerk. "We don't talk about that. Especially in my profession." I looked around to see if we were again being followed, but we appeared to be unwatched.

All around us people were walking purposefully. On the main street lining the square, hundreds of identical black bicycles flowed by, ridden by people bundled into drab-colored jackets and baggy pants; every few minutes a car or bus would go by. It was a striking contrast to the cities of Germany, where I then lived, where a busy city street meant a perpetual clot of cars, and bicycles were rare except in the countryside.

This brought my mind back to the differences between East and West. I wondered if this young father and mother had considered infanticide when their daughter was born, as so many other Chinese parents had. I wondered what it would be like to be told by my government I could have no more than one child. I knew I would have rebelled, but in China rebellion meant consignment to labor or re-education camps, harsh and brutal places that often made death seem like a pleasant option.

"Do you think a government should have the power to dictate how many children a family can have?" I said, wanting to close the issue in my mind.

Dr. Wu pushed his hands deeper into the pockets of his pea-green overcoat and sighed loudly, as if he'd felt a pain deep in his stomach. "You may not remember 1960," he said after a few steps. "There was a terrible famine, and many people died."

I had a vague memory of the famine in China in 1959–1961, when 30 million people died of starvation. They were abstractions to me, however: I was ten years old at the time. I'd never heard any details, seen any pictures, or met anybody who'd lived through it.

"I've heard of that time," I said.

"China is now over a billion people," he said, his steps heavy on the damp cement. "That's a fifth of the world's population, just in our country, and twice the population we had at the time of the famines. And today, now, we are straining to feed our billion people. Something must be done. You understand?"

"Yes," I said. "But what about simply increasing agricultural output? Most people in this country are farming with oxen or using their wives to pull their plows, when you're sitting on reserves of oil and coal. In my travels around China I've seen hundreds of people in the fields, but only one or two tractors."

"No," he said. "That would only make things worse. Produce

more food and there will be more people. There are already too many people in China, which is why we are so poor."

"But you have vast natural resources. . . ."

"We have a destiny," he said, standing straighter. A new tone entered his voice, one of command and authority, as if he were speaking to a less intelligent underling, giving orders to a nurse or an orderly as I'd heard him do in the hospital. "China will not follow the mistake of the West. We have learned from our past."

"Mistake?"

"If the winter is coming and you have stored enough food to make it through that winter with your family, would you allow your children to eat all the food in the first month of the winter?"

"No, of course not."

"China will not make this mistake with oil," he said in flat tones. "We will build hydroelectric generators, and use some of our forest and coal, but we must keep our stock for the winter."

"When is the winter?"

"When America and Europe and the Middle East have run out of oil. We will survive. There are even some who say we will then dominate the world, that that time will signal the Third Chinese Empire." He shivered and looked at the ground. "But that is dangerous talk."

"Like the Third Reich?" I said, and instantly regretted having said it.

He looked at me as if he were judging an enemy on the battlefield. "There are nationalists in China as there are in every country."

I felt uncomfortable with his sharp tone, and so tried to move away from the subject. "Are you familiar with the studies in the United States about the effect of birth order on psychology?"

He nodded as we approached the museum, a huge building that houses the most ancient bits of known technology to ever have been produced by any civilization. "Yes, I am familiar with those studies," he said.

"Do you have siblings?"

"I have two older brothers and a younger sister," he said, his tone cautious. "What about you?"

"I have three younger brothers," I said.

"Born to leadership."

"What?"

"The oldest in such a large family is born to leadership," he said. "That's what your studies found. The oldest learns to manage those under him."

I shrugged, not wanting to get into psychoanalyzing either one of us. "But we both have siblings," I said. "You know how only-children, those born into a family where they are the only child in the family, have a unique psychology?"

"Yes," he said. "Depending on how they are raised, of course. But often they are quite dominating, since the entire attention of the family has been lavished on them for their entire lives. They have a unique psychological profile."

We climbed the stairs to the museum and walked inside. The huge brick and marble building was not heated, and our voices mingled with a hundred other echoed tones as people walked along the cold stone floors, viewed the displays, and chatted softly. We stopped in front of a water-driven clock that was five thousand years old. Made of brass and stone, it stood four feet tall and three feet across. Water dripped through it, moving levers and filling tanks, the levels showing the hour of the day.

"What will China be like when all of the leaders, all of the managers, all of the population are only-children?" I said softly.

He looked at the clock for a long moment. "It will be winter, and we will still have our season's supply, although the rest of the world will have consumed theirs." He turned his dark brown eyes toward me, looking directly into mine. "It will be the time of our destiny."

Who will feed China?

That was 1986, two years before Worldwatch Institute co-founder and author Lester R. Brown would begin to talk and write about the convergence he saw coming between China's growing food needs and her leveling-off ability to increase food production. In 1994, when he published his article "Who Will Feed China?" pointing out how the trend of industrialization in China would inevitably lead to her becoming a food-importing country, he was attacked by no less than the government of China itself as being an alarmist.[2]

However, in 1995 as the inevitable industrial trend of replacing farmland with factories and roads continued, and population mounted, China changed her tune. That year, the most populous nation on Earth had to import food to feed herself, and it sent shocks through the world's grain markets.

Over the next 20 years, Brown predicts, China's need for imported grain will grow from a few million tons to over 200 million, and perhaps as much as 300 million tons. Yet, according to the U.S. Department of Agriculture, grain exports of *all the food-exporting countries of the entire world* in 1994 were less than 230 million tons. And those exports are helping to feed more than a hundred nations that have become food importers: only a few dozen export, and only Canada and the United States export grain in any significant quantity.

Further, the ability of Canada and the United States to grow enough food for export is tenuously balanced on the shifting whims of weather. In 1988, which was at that time the hottest year in history, droughts in the American Midwest cut Canadian grain production by 37 percent and U.S. grain production by 30 percent, causing the United States to consume more grain than it produced for the first time in over three hundred years.

When China becomes hungry over the next few years, her need for food will rock world food prices, according to Brown and others. As prices rise worldwide, those countries unable to cover the rising cost of food will tip over into famine as China, now a goods-exporting powerhouse, can raise the money to pay for her food.

Food may well become the commodity that's scarce long before oil dries up. Unfortunately, however, so few people in the most affluent parts of the world—who have the financial and political power to be most effective in doing something about this—have any real understanding of the situation.

In September 1997, *National Geographic* magazine ran an article about another aspect of how far China will go to survive: the Three Gorges Dam project on the Yangtze. This 600-foot-high dam, which has now been built, created a reservoir 370 miles long (the size of Lake Superior) with a series of locks to bring ocean commerce far inland. Inside the dam are 26 of the world's biggest turbines. At 400 tons each, they generate 18,200 megawatts of electricity, equivalent to the output of 18 nuclear power plants, all in one dam.

That's the equivalent of 90 million barrels of oil per year. It's also the equivalent of a 10,000-square-mile solar panel operating at 100 percent efficiency. Critics cite the disastrous effects (and perhaps folly) of trying to dam such a massive and powerful river. Some say it'll turn the reservoir into an open sewer. And, *National Geographic* pointed out, a "Canadian environmental group cites the poisoning they say will happen from industrial toxins leaching out of drowned factories: arsenic, cyanide, methylmercury."

And well it must. A report by the British power giant PowerGen, titled "Energy 2020," projected that within two decades the two largest consumers of power on the planet would be the United States (which, with 6 percent of the world's population today consumes about 25 percent of the world's energy) and China, where, they say, demand is growing at twice the rate it ever has in the United States.

Clearly, the Chinese are willing to pay almost any price not to be dependents in the coming decades.

The Last Hours of (Cheap, Clean) Water

 Only modern "advanced" cultures, driven by acquisition and convinced of their supremacy over Nature, have failed to revere water. The consequences are evident in every quarter of the globe: parched deserts and cities, destroyed wetlands, contaminated waterways, and dying children and animals.

—MAUDE BARLOW and TONY CLARKE in *Blue Gold: The Fight to Stop the Corporate Theft of the World's Water*

In the late-nineteenth and early-twentieth century the landscape of Pennsylvania, New Jersey, New York, Ohio, and other eastern and midwestern states were dotted with oil wells. The Pennzoil Company, for example, came out of that era as Pennsylvania oil was considered very high quality. But today Pennzoil's petroleum comes mostly from overseas, because the oil wells of Pennsylvania and the rest of the eastern United States were pumped dry.

As those wells ran out, exploration moved west. Oil was found in Oklahoma, California, and Texas, so plentiful that it bubbled from the ground. Much of that has now gone, too, and the oil that once flowed from Alaska—providing a subsidy to Alaskan citizens—has also dried up, with only the protected regions of the state still holding a small amount of oil. As American oil ran low, the search for fuel moved to the Middle East, the former Soviet states to the north, and Africa to the south.

For the moment, we can ship oil from overseas to feed the appetite of American industry and drivers. Trying to keep unstable oil-producing regions under the American thumb may make for messy politics and questionable morality, but it can be done, at least for another decade or two or three.

It's not so easy, though, with water.

Collapsing aquifers

Aquifers are pools of underground water, and they take two major forms: confined and unconfined. These are either replenished rapidly or slowly or not replenished.

Rapidly replenished aquifers are a variation on filtered ground water. Rain and melted snow sink into the ground and slowly work their way through layers of earth and underground rock. They eventually reach layers of sand or porous rock, where they either accumulate or begin to flow like slow underground rivers. The water in those close to the surface may be only a few years old (typically from two or three to a hundred years), whereas deep replenished aquifers may have water thousands of years old.

Some replenished aquifers—such as the ones most rural home wells tap into—replenish at a rate commensurate with the depletion rate. They're fragile in that drought can flip them off in as little as a year or two (as residents of Maine found out in 2001 when wells all across the state ran dry), but at least they're replenished when the rains fall.

Slowly replenished aquifers are more like oil wells. Consider, for example, the Ogallala Aquifer, which stretches from Texas to South Dakota, covering 174,000 square miles and holding enough water to fill Lake Huron.

The Rocky Mountains erupted 80 million years ago, and over the following 60 million years erosion and tree roots created a vast landscape of soil and sand. Over the past 20 million or so years this sand layer has become submerged sandstone and gravel, trapping water that slowly seeps into it from the mostly arid lands above. The Ogallala Aquifer averages about 200 feet thick, and during the 40-year period from 1940 to 1980 it dropped an average of 10 feet, losing as much as 100 feet in some parts of Texas. The recharge rate is so slow that much of the water being pumped out of the aquifer is more than 10,000 years old, and when it's drained it'll take 10,000 to 15,000 years to recharge.

Right now the amount of water being pulled out of the aquifer, both for residential use and as irrigation for agriculture, is greater than the total flow of the Colorado River. Estimates vary, but there's a broad consensus that the aquifer will pump dry (or at least low enough to be unusable) at current rates of usage sometime in the next 30 to 50 years.

Over 40 percent of the grain grown in the U.S. and about a third of our cotton is irrigated by the Ogallala, and a quarter of all the feed grains exported by the United States are grown on its water. In total, over 14 million acres of cropland draw their sustenance from the aquifer—land that will become unproductive when it runs dry. The consequences will be substantial, both economically for the U.S. and in terms of world food, since U.S. grain exports account for the survival of millions of people in the world's poorest nations.

As the U.S. Department of Agriculture writes on its website,

> Areas impacted by overdraft of ground-water resources include the High Plains of Texas, the Ogallala aquifer is near depletion. The state of Texas has lost 1.435 million acres of irrigated cultivated cropland over the period, 1982–1997. Most of the loss is due to dwindling ground-water supplies from the Ogallala aquifer. Aquifer level declines have ranged from 50–100 feet since 1980 with saturated thickness reductions of 50%. Depth to water is highly variable but commonly exceeds 100 feet. Well yields are down from 1000 gpm to 250 gpm and less in some areas. At current rates of pumpage, current irrigated acreage is predicted to drop 50% by 2030.[1]

The USDA also points out the following:

> Some aquifers have been permanently damaged because full recharge of depleted aquifer storage will not be possible where compaction and subsidence have occurred. Mining of ground water reserves to sustain agricultural production is temporary.

Declining water levels will increase pumping costs and aquifer depletion will decrease well yields to the point where aquifers will be abandoned for irrigation usage. Long-term solutions must be investigated and evaluated today to maintain U.S. food and fiber production for tomorrow.

Unfortunately, funding for long-term planning was cut during the Bush administration.

Another midwestern aquifer demonstrates the problems of over-pumping. The U.S. Army Corps of Engineers points out that "the Grand Prairie's Alluvial Aquifer will be too small for commercial use by 2015." This aquifer underlies the Grand Prairie of Arkansas and, the Corps says, "Currently, more than 90 percent of the water needed for crop irrigation is being withdrawn from the Grand Prairie's Alluvial Aquifer. A main source of agricultural water since 1904, the Alluvial Aquifer annually provides millions of gallons of water but is quickly being depleted. In fact, the water table in the aquifer has been declining at the rate of about one foot per year and since 1915, faster than natural recharge."[2]

It points out that if you drilled into the Alluvial Aquifer in 1937, you'd have found water just 50 feet below the surface. Today you have to drill at least 105 feet down. As the aquifer has been drained, the land of the Grand Prairie has collapsed in a way that's easily visible from the air. "This de-watering of the Alluvial Aquifer," the Corps notes, "has caused a cone of depression in it that resembles a long trough, which centers around Stuttgart, Arkansas and extends northward to Hazen, Carlisle and Lonoke. Recent studies have shown that the cone of depression is lengthening and now stretches nearly to England, some 35 miles northwest of Stuttgart."

Another smaller aquifer, the Sparta Aquifer, located below the Alluvial Aquifer, has been suggested as an alternative, at least for drinking water. "However, the Sparta Aquifer does not have the capacity, has extremely slow recharge, and is expensive to pump," the Corps says, and "The Sparta Aquifer is already being tapped for uses other than drinking water and its level has been declining since 1986 at the rate of about one-foot per year."

Most other major aquifers in the United States are also at risk, reflecting water problems around the world.

In Florida, for example, the massive Florida Aquifer is being so rapidly depleted by the state's growing population that they've begun injecting treated human sewage into it to try to recharge/refill it before it runs dry. The hope is that the gravel in the aquifer will clean up the sewage before the water from the state's toilets works it way back up to the drinking-water-drawing wells. The injection is done using what's called a "Class 1 Underground Injection Control (UIC)" well, similar to those used in other parts of the U.S. to pump toxic waste into the ground. In this case, 400 million gallons of municipally treated human waste are injected into the aquifer every day.

Unfortunately, according to the Environmental News Service, nobody yet knows if this scheme will be safe.[3] Will human viruses survive being pumped back up and chlorinated? Or, more problematic, how will the tons of unmetabolized drugs that go from Floridians' kidneys into toilets every day—drugs ranging from antibiotics to blood pressure medications to tranquilizers—affect people when the drugs one day show up in Florida's drinking water? Nobody knows.

Following the lead of George W. Bush, who made many environmental compliance rules in Texas voluntary for big corporate polluters (leading to Texas having the worst toxic pollution problem and the worst air in the U.S.), Jeb Bush's Florida has instituted a similar program. ENS reports, "While Class 1 well permit operators are required to monitor underground sources of drinking water, the program is voluntary and his [the DEP UIC monitor's] office doesn't have the funding to collect the data. EPA Region 4 administrators admit they are relying solely on the limited DEP voluntary UIC monitoring program to assess the health impact of Class 1 waste entering the Floridan."

Other aquifers around the United States are either recharged with purchased river water or are running dry. For example, while the community of Palm Springs, California, can afford to recharge its aquifer with water diverted from the Colorado River, the California town of Borrego Springs has no such luxury. The *San Diego Union Tribune* reported on July 6, 2000, "It's called Borrego Springs, but there are no springs. They dried up long ago due to extensive pumping of the ground-water aquifer beneath the baking desert floor." The article goes on to point out how in Colorado, underground water is considered part of the commons and thus regulated by the government, which represents all residents in the state. In California, however, water is considered the property of whoever drills a well and finds it, so the large citrus-growing operations around Borrego Springs have drained so much water out of the area's aquifer that the "underground water level in wells around Borrego has fallen 30 feet in the last 13 years, causing desert trees that once dipped their roots into it to die of thirst."

Another city that's facing a crisis because of aquifer depletion is Mexico City. Built over a confined aquifer trapped in the upper layers of an extinct volcano, Mexico City has fallen 30 feet in the past century because of ground settling into sand once thick with water. The unevenness of the sinking has created havoc throughout the city, cracking ancient buildings, bursting over 40,000 water and sewer pipes a year, and turning the city's once-flat rapid transit rail system into a roller coaster. A water pipe drilled into the aquifer and at ground level in 1934 now hangs 26 feet in the air as the city has dropped around it, and the city has had to invest in expensive pumping equipment to push liquid sewage up to a level where it will drain out of a city sewer system that was once well underground.

Similar aquifer depletion is causing Venice to collapse, and affecting hundreds of other cities all around the world. For example, the North China Plain (including Beijing) sits over a deep aquifer that is in deep trouble. A World Bank report cited by Lester Brown in a March 13, 2003, paper titled "World Creating Food Bubble Economy Based on Unsustainable Use of Water" says, "Anecdotal evidence suggest that deep wells [drilled] around Beijing now have to reach 1,000 meters [more than half a mile] to tap fresh water, adding dramatically to the cost of supply." Brown notes in his paper for the *Earth Policy Institute* that "In unusually strong language for the Bank, the report forecasts 'catastrophic consequences for future generations' unless water use and supply can quickly be brought back into balance." Brown cites aquifers in India that are dropping by up to a meter a year, and says that, according to the World Wildlife Fund, in Pakistan the aquifer supplying the provincial capitol of Quetta will run dry "within 15 years."

Aquifer depletion is hitting Yemen hard, with aquifers "falling by roughly 2 meters a year," according to Brown's report, and food shortages will begin all around the world within the next decade or two as a result of the loss of irrigation water.

Those shortages will be as politically destabilizing and threatening to human (and other) life as are the oil shortages that will occur around the same time.

River waters at risk

On April 28, 2003, the *Wall Street Journal* ran an article by Peter Waldman that had come out of some excellent investigative reporting. Perchlorate—a toxic chemical used in rocket fuel—has come to contaminate drinking and river waters in more than 20 states in the United States, and just before publication of the *Journal* article was found to contaminate, in relatively high concentrations, lettuce grown with Colorado River water. It turns out that the lettuce concentrates the chemical in its leaves, so even if the water irrigating it is contaminated below "safe" levels, the lettuce can end up with much higher amounts.

For example, the EPA has defined perchlorate concentrations above one part per billion (1 ppb) to pose "dangers to human health, particularly to infant development," according to the *Journal*. But among lettuce tested, "Four of the 22 samples tested were found to contain perchlorate in excess of 30 parts per billion, with the highest—'mixed organic baby greens'—registering 121 ppb." The *Journal* reported that the EPA's studies showed that lettuce concentrates perchlorate by a factor of 17 to 28, "meaning the outer leaves contained 17 to 28 times more perchlorate in them than did the water used to irrigate the plants."

But that's only the beginning of the story.

When the EPA did its investigation and submitted it for publication in a peer-reviewed journal, it came to the attention of the defense industry. It apparently sprang into action, calling up those beholden to it in the government. Just a few months earlier, the White House had proposed that the military and defense contractors be exempted from liability for perchlorate cleanups, and when the lettuce studies began to leak out, the George W. Bush administration really went to work.

The White House ordered its own EPA not to publicly discuss the lettuce and perchlorate issue, going so far as to impose a gag order on the agency. The EPA is not allowed to discuss its test results with any members of the press, or to publish the research it was preparing for publication before the gag order. At the same time, Senator James Inhofe (R.-Oklahoma, and not a scientist) began strongly arguing with the EPA that it should loosen its guidelines for perchlorate contamination. He agreed with several defense contractors who said levels of 70 ppb to 200 ppb in drinking water should be considered acceptable, instead of the current threshold of 1 ppb.

This is just one very small example—there are literally thousands—of how corporate money has corrupted politics to the point where government, which in a democratic republic is supposed to represent "We, The People," has left a nation's water supplies and its citizens at risk.

Drinking water unsafe?

People concerned about the contamination of tap water have turned to bottled water in such numbers around the world that it has become one of the most popular and profitable of the consumer beverages in the market. Yet a study published in *Nature* on April 8, 2002, calls into doubt the safety of bottled water.

"In 11 of 29 European brands of bottled mineral water," the article's author, Natasha McDowell, reported, "Christian Beuret and colleagues of the Cantonal Food Laboratory in Solothurn found signs of the virus that causes more than 90% of the world's stomach upsets. The virus is called Norwalk-like virus or NLV." Viruses in the Norwalk family were responsible for the persistent and recurrent epidemics of diarrhea on cruise ships that got so much press coverage in 2001.

In the search for the planet's dwindling supplies of fresh water, a

new industry has emerged in the past decade: mining icebergs. Floating mining operations as well as on-land operations in Iceland have
been set up, leading to iceberg bottled water, iceberg vodka, and iceberg beers. Even this, though, is a limited asset resource—as global
warming increases, icebergs around the world are shrinking instead of
growing.

At the same time, drugs are showing up in the planet's water supplies in startling levels, and their effects are not limited to the humans
who first took them. It first came to the attention of the scientific
world in 1998 when Swiss chemists were screening lake water, looking for the herbicide mecoprop. Instead, they found clofibric acid, a
prescription drug used to lower cholesterol . . . that isn't even manufactured in Switzerland. Looking for a wider range of substances, they
discovered lipid regulators phenazone and fenofibrate, analgesics
ibuprofen and diclofenac, and a veritable cocktail of other prescription drugs ranging from chemotherapy agents to hormones to
antibiotics to beta-blockers, and a smattering of antiseptics, anticon-
vulsants, and X-ray contrast agents.

These results stimulated an American researcher to look at the
waters flowing into Lake Mead (Nevada and Arizona), where he
found a similar cocktail, along with the female hormone estradiol—
at levels up to 20 ppt (parts per trillion) which, according to a report
in *Science News,* are "a concentration that can cause some male fish to
produce an egg-making protein normally seen only in reproductive
females."[4] Another concern, cited by Stuart Levy, the director of the
Center for Adaptation Genetics and Drug Resistance at Tufts University in Boston, is antibiotics in the waters that "may be present at
levels of consequence for bacteria—levels that could not only alter
the ecology of the environment but also give rise to antibiotic
resistance."

Two years later, the April 1, 2000, *Science News* reported in an
article titled "More Waters Test Positive for Drugs" that "the
cholesterol-lowering drug clofibric acid turned up in a groundwater
reservoir being tapped to meet the Phoenix community's thirst. The
drug had entered with treated sewage, which the city had been using
to replenish the acquifer." The article's author, J. Raloff, notes

groundwaters around the world are now being found tainted with a broad variety of drugs, some clearly from humans and others from factory farming operations, since "an estimated 40 percent of antibiotics produced in the United States is fed to livestock as growth enhancers."

Up to 90 percent of those drugs are excreted unaltered or only slightly altered. Raloff notes that "the biggest risks face aquatic life—which may be bathed from cradle to grave in a solution of drugs of increasing concentration and potency." And Raloff cited Stanford University's David Epel, who "expressed special concern about new drugs called efflux-pump inhibitors. Designed to keep microbes from ejecting the antibiotics intended to slay them, efflux-pump inhibitors also impede the cellular pumps that nearly all animals use to get rid of toxicants, [Epel] says. If pump-inhibiting drugs enter the aquatic environment, Epel worries that they might render wildlife vulnerable to concentrations of pollution that had previously been innocuous."

A June 29, 2002, article in *Science News* titled "Pharm Pollution" pointed out that farm "animals excrete up to 90 percent of the tetracycline administered," and that while humans consume 4.5 million pounds of antibiotics annually, and 2 million pounds are used to treat sick animals, fully 27.5 million pounds are fed routinely to animals every year in the U.S. alone "for growth promotion."

Over-the-counter and prescription drugs that have flooded the environment with our trillions of gallons of wastewater are altering ecosystems everywhere they go. Marine animals, plants, wild animals, insects, trees, and even bacteria have been documented to experience biological shock from these persistent and highly biologically active compounds.

As Gannett science writer Traci Watson pointed out in an article titled "Frogs Rapidly Vanishing Across U.S.," "The nation's amphibians—frogs, toads and salamanders—are vanishing at an alarming rate, their music ending in many parts of the American landscape." USGS (U.S. Geological Survey) herpetologist Bruce Bury pointed out that "We've done so much to our water systems that anything tied to water is in bad shape these days." Gopher frogs in the south-eastern United States, boreal toads that used to present dangers to drivers in Colorado, and "every species of frog and toad in Yosemite National Park" are all disappearing. Concludes Watson, "Silent, frogless nights are common in other parts of the [developed] world, too. In Costa Rica, the famous and highly protected golden toad, named for its bright skin, has become extinct. In the 1980s, 8 of 13 frog species in the Brazilian preserve vanished. In Australia, the gastric-brooding frog, named for its habit of incubating its young in its stomach, broods no longer."[5]

As Stephen Harrod Buhner points out in his brilliant book *The Lost Language of Plants:*

> Under the pressure of pharmaceuticals, genotypes throughout ecosystems—bacteria insects, viruses, plants, and more—are going fluid and reassembling themselves in order to reestablish system homeostasis. The more intensely human chemicals act as environmental stressors on the system, the greater the pressure on genotypes in response. This is the primary reason it takes more and more chemicals each year to produce the same level of results in agriculture and medicine. Eventually the amount of chemicals needed will surpass energy supply and the system will contract rapidly in order to reestablish homeostasis. The amount of pharmaceuticals being produced (especially when combined with agrochemical production) is enough to interfere with the homeostatic balance of plant communities, ecosystems, biomes, and Earth itself. In many instances these chemistries combine together with each other, and sometimes with plant chemistries, in synergistic ways that are not predictable and that produce magnified impacts on ecosystems.[6]

Water for profit

For most of the history of democratic governments, water was considered part of the commons—the things we all use that are held on our behalf in common trust by governments answerable to "We, The

People." If we didn't like how our water was being used or treated, we had the option of complaining to our elected officials and even voting out of office those who didn't take seriously the sacred trust of holding office and protecting the common wealth.

In the 1930s, however, Benito Mussolini came up with a new idea. He initially called it "corporatism," thinking he'd invented something new, and suggested that governments could become immeasurably strengthened if they worked hand-in-glove with the most powerful corporate powers in a nation. The corporations would take over many of the traditional functions of government, handling them with efficiency and increasing the revenues of the corporate stockholders (which would help the economy), and the people in power (then Mussolini) would have a greater ability to hold power because the corporations controlling the media and the industrial infrastructure would, of course, support them. Mussolini envisioned the bundle of sticks bound together with a rope, with an ax sticking out of the top, that was the symbol of the ancient Roman emperors as the logo for his new governmental style, and renamed it fascism after the bundle, which was known in Latin as *fascie*.

Franco in Spain and Hitler in Germany both saw fascism as a great way to consolidate and enlarge their power, and to ensure that they'd never, in their lifetimes, face serious political opposition. They embraced fascism, and the rest is history.

I share this with you as prelude to what's been happening in the United States and, increasingly, around the world, ever since the "Reagan/Bush Revolution" of the 1980s. Reagan cut taxes to run up national debt, loosened regulations, stopped enforcing antitrust laws, and embraced corporations in a way never before seen in the United States. The pace was slowed a bit by the election of Bill Clinton, but then Clinton amazed democracy advocates when he pushed through the U.S. Congress legislation—during a lame duck session just before the Christmas recess when lawmakers were anxious to get home—that elevated corporations to the status of nation-states in many ways. The beginnings of Mussolini's vision were firmly seated in the U.S. and around the world by passage of WTO/GATT and NAFTA, followed by George W. Bush's dramatic embracing of corporate executives to fill his top administration spots, corporate money to finance his campaign, and massive interlocking programs with corporations and government.

In this context, programs to "privatize" water begin to make sense. It provides another way for corporations to profit from government activity (and reward the politicians who set up the schemes), while shifting what may well be the world's most valuable single resource into the hands of a very wealthy few.

Consider: every 20 years global consumption of water by humans is doubling, while at the time of this writing the U.N. says that over a billion people lack access to fresh drinking water (other sources suggest if you insert the word "safe" before "fresh," the number may be closer to three billion people). Since only about one-half of 1 percent of the planet's water is drinkable and accessible to public water systems, at this rate of consumption, the human need for fresh water will outstrip the entire planet's replenishable supply by 2025—even before we run out of oil.

This fact hasn't been lost on some of the world's largest corporations, or on the World Bank, which has taken on the role of patron saint of corporate-state partnerships. Water supplies from Cochabama, Bolivia, to San Francisco to Atlanta to England to India are being privatized at a rate so fast as to be dizzying. Jeb Bush cut a deal with Enron to privatize large parts of the Everglades, a British transnational is cutting deals with companies all over California, and almost every water supply on the planet is being raided by corporations ranging in size from the world's largest to small startup outfits with political connections.

It's a gold rush that makes California in 1849 look like a picnic outing, and there is absolutely no discrimination going on: First World water supplies (and the politicians who are supposed to hold them for the public) are just as up for grabs as are those in the Third World. There's even a nonprofit that claims to "watch" privatization trends—and is prominent on the Internet—but is actually an industry front group.

The rationalization for privatization (of water and everything else) offered by industry and corrupted politicians is that corporations are more interested in efficiency than are government bureau-crats. History and experience, of course, prove they're right, and so-called "conservatives" (Dwight Eisenhower wouldn't recognize them) try to stop the discussion right there.

But what's unspoken is that while corporations are often more efficient than government at performing specific tasks, that doesn't mean that the task is done at a lower cost to the end user. Corpora-tions, after all, must earn profits and pay dividends to their stock-holders. They have fancy corporate headquarters and corporate cars and corporate jets to pay for. They lavish huge salaries, bonuses, and perks on their top executives. And they spend millions on lobbying, political campaign contributions, and advertising/PR/marketing.

Because governments don't have any of these expenses, the Medicare program in the United States (to use one example) is able to consistently deliver health care to its users (as of this writing in 2003) at around 3 percent of revenues. In other words, for every $100 spent for health care through the Medicare program, $3 sticks to the fingers of the administrators to cover their salaries, offices, and other costs. On the other hand, while corporate health-care pro-viders can deliver services for the slightly more efficient $2.75 (var-ies by region) per $100, the actual cost to companies and employers who use private health insurance, HMOs, and PPOs is between 14 and 20 percent. Why the increased costs? The health insurance com-panies must pay part of their revenues to cover all those corporate-associated costs mentioned above. So, while they can argue that they're more efficient than government, the reality is that the cost to the consumer is higher.

The other problems associated with privatization include trans-parency and lack of accountability. The George W. Bush administra-tion used privatization skillfully to move many police and military functions to private corporations so the actions of the administration would no longer be subject to public scrutiny. Freedom of Informa-tion Act filings can be ignored by corporations, who claim that a bizarre interpretation of an 1886 Supreme Court ruling turned them into "persons" under the constitution and endowed them with human rights, including the Fourth Amendment right to privacy.

Corporations are also not accountable to their customers, at least when they have monopoly control of something those customers need. (This is where the whole "competition breeds good corporate behavior" argument breaks down.) In fact, by law a corporation is accountable to only one entity—its shareholders. And the shareholders will always demand profits in their hands over all other considerations.

Thus, when citizens in a democracy are unhappy with how their water supply is handled by government, they know the agency and have the names of the elected or appointed officials to petition for redress or repair. They can even vote out of office those who appoint the officials if they don't get satisfaction. On the other hand, as Michael Moore so brilliantly demonstrated in his movie *Roger and Me* about his attempts to speak with G.M. president Roger Smith on behalf of the citizens of Flint, Michigan, corporations can—and regularly do—confront consumers with locked doors, endless voice mail loops, and low-level corporate bureaucrats whose main job is to keep the riffraff away from the executive offices.

When the citizens of a small city in Bolivia encountered this sort of treatment from one of the world's largest corporations that had privatized their water supply, they revolted. The local government was forced to break the privatization deal and tried handing over the water system to the protestors. The corporation, however, incorporated part of itself in the Netherlands and used that venue to sue the local Bolivian government for millions under a bilateral treaty Bolivia has with the Netherlands. They also threatened to sue journalists reporting on the story, cutting down on media coverage around the world (the news is now also a for-profit business in most parts of the world, and lawsuits make it less profitable so when corporations sue they normally get their way). The case is still in court as of this writing, but has cast a chill over citizen efforts to reclaim privatized water all around the world.

Deforesting, Fighting for Fuel, and the Rise and Fall of Empires

 An empire is an immense egotism.

—RALPH WALDO EMERSON (1803–1882)

One of the most historically compelling empires of all time is that of the Sumerians. A friend of mine in Atlanta can tell you all about this ancient civilization. His dictionary collection contains a stone fragment that was carved between six and eight thousand years ago in the Sumerian kingdom of Uruk, in Mesopotamia, which is now known as Syria, Iraq, and Lebanon.

"Most people don't even remember the early Sumerians," Tom told me as he lovingly handed me his Sumerian dictionary fragment. "But they were the earliest fathers of our way of living, what we call Western Civilization."

According to the *Epic of Gilgamesh,* the oldest written story in the world, one of the first kings of the earliest Sumerian civilization (the Uruks) was a man named Gilgamesh. He was the first mortal to defy the forest god, Humbaba, who had been entrusted by the chief Sumerian deity, Enlil, to protect the cedar forests of Lebanon from mankind.

King Gilgamesh wanted to build a great city, Uruk, to immortalize his contribution to Sumerian civilization. So he and his loggers rebelled against Humbaba and began to cut the forests, which then stretched from Jordan to the sea in Lebanon. The story ends with Gilgamesh decapitating Humbaba and infuriating the god of gods,

Enlil. Enlil then avenges the death of Humbaba by making the water in his kingdom undrinkable and the fields barren—thus killing off Gilgamesh and his people.

Along with its other distinctive qualities, the *Epic of Gilgamesh* is the earliest recorded story of downstream siltation and desertification caused by the extensive destruction of forestlands. Lebanon went from more than 90 percent forest (the famous Cedars of Lebanon) to less than 7 percent over a 1,500-year period, causing downwind rainfall to decrease by 80 percent. As we have seen, trees and their roots are an important part of the water cycle. As a result, millions of acres of land in the Fertile Crescent area turned to desert or scrubland, and remain relatively barren to this day—fertile no more.

The staple food of the Mesopotamians was barley, but over a period of several hundred years of continuous growth of barley on irrigated land, the land became exhausted and had such high levels of salt (carried in by the irrigation water) that it would no longer grow crops. At the same time, because of the rapid destruction of the forests, wood had become such a precious commodity that it was equal in value to some gem and mineral ores: neighboring countries were conquered for their wood supplies, as well as to get fertile land to grow barley. Vast areas of timberland along the Euphrates and Tigris rivers were cut bare, increasing the siltation of their irrigation canals and cropland and further decreasing downwind rainfall.

The result of this local climatic change more than five thousand years ago was widespread famine. The collapse of the last Mesopotamian empire happened around four thousand years ago, and the records they left behind show that only at the very end of their empire did they realize how they had destroyed their precious source of food and fuel by razing their forests and despoiling the rest of their environment. For thousands of years they "knew" that their way of life was fine. But although things looked good at the time, they didn't realize it wasn't sustainable: it worked only as long as they had other people's lands to conquer. Once they ran out of neighbors, their decline was sudden and devastating, just like a Ponzi scheme.

* * *

The collapse of the Mesopotamian empire paved the way for the rise of Greece as a "worldwide" empire in the late Bronze Age. Between 200 B.C. and 1500 B.C., the Greeks adopted widespread agricultural practices similar to the Mesopotamians' system. By the thirteenth century B.C., this increased food supply caused the Greeks to begin clearing huge tracts of their forest to provide living space, fuel, and cropland for their ever-growing population. They also used tens of thousands of acres of forest wood to feed the bronze furnaces for which they were famous.

The decline of their civilization is linked in the historical record to their population outstripping their available fuel: wood. By 600 B.C., most of Greece was an environmental wasteland, with denuded hillsides eroding into over-silted rivers and irrigated cropland collapsing under increasing levels of accumulated irrigation-salts and nutrient exhaustion. A bounty was offered to farmers to grow olive trees on the hillsides, as the desperate Greeks found that only olive trees could grow and hold to the fragile, steep slopes. But it was all too late. As Plato wrote in his *Critias:*

> What now remains compared with what then existed is like the skeleton of a sick man, all the fat and soft earth having wasted away, and only the bare framework of the land being left.

It has, indeed, happened before.

* * *

The collapse of Greece was followed by the rise of the Roman Empire.

Rome had its own needs for wood: by 200 B.C. the forests of what we now call Italy were all but wiped out to meet the Roman needs for fuel and shelter, to warm the public baths, and to smelt metals. Large quantities of wood were necessary to smelt silver from ore, refine the metal, and mint it into the coins that were the basis of Rome's monetary system. So when Italy's forests were exhausted around the first century A.D., the repeated doublings in the cost of wood to smelt silver caused a monetary crisis, the first huge crack in the Roman Empire.

About that same time, the productivity of Roman croplands

began to collapse because of siltation, salination, soil exhaustion, and decreased rainfall from the loss of upwind forest. The food shortages threatened the stability of the Roman Empire. This led Rome's leaders to build a fleet of 60 wooden ships to conquer the nearby Mediterranean countries, extending the empire in its last days as they reached out across the known world for minerals, food, and wood. Ultimately, Rome's watershed destruction, deforestation, depleted soil, and booming population led to widespread famines, resulting in the collapse of the Roman Empire.

Even mighty Rome was not sustainable, did not live within its means—even after conquering half the known world. And of course this is only a small snapshot: hundreds of other Younger Culture experiments have erupted, wiped out their resource base, and died off, from the Egyptians to the citizens of Ur, the dynasties of China to the vanished civilizations of pre-Colombian South America.

* * *

When America started on the road to becoming the most powerful force in its world, her main fuel was wood. It was the fuel and heat source for George Washington's army, and remained our nation's primary source of fuel, heat, and building material through the time of the Civil War, when coal began to be used extensively.

As mentioned earlier, the discovery of oil in Pennsylvania just after the Civil War dramatically increased the human race's ability to grow and feed the global population. And that leads us into a situation that's uncomfortably similar to Mesopotamia, Greece, and Rome: now that we have all these people dependent on a particular fuel, what happens when it runs low?

Can we save our civilization with alternatives to oil?

For many years, people have been pointing out that we will eventually—and relatively soon—run out of oil. They've proposed a variety of "alternative" or "renewable" energy sources to replace oil. Since the goal of all of these is to reduce our dependence on oil—thus averting an energy-collapse-caused crumbling of our civilization, like the Sumerians'—they are worthy of examination.

A big economic hurdle: artificially low oil prices discourage investment

When considering any of these, however, we have to keep in mind why there has been so little emphasis, so little sense of urgency among governments, about the development of alternatives to fossil fuels.

The primary reason is that because we're pumping oil worldwide so aggressively, the oil we have right now is cheaper (adjusted for inflation) than it ever has been in the history of the world. A pint of bottled water in the store can cost four times more than a pint of gasoline—even though the gasoline was pumped up from the ground eight thousand miles away as crude oil, shipped here, refined, and transported across the country to your gas station.

In part, the low price is because Americans are paying for the oil with their tax dollars—annual subsidies, grants, gifts, and tax breaks to the industry typically run between $20 and $84 billion dollars, according to several studies documented by Greenpeace International, the Media Awareness Project, and other watchdog groups. (Those subsidies rose substantially above the $84 billion figure when George W. Bush took office in 2001.)[1] Big oil donates hundreds of millions to politicians in the U.S. and elsewhere, and in return receives billions in subsidies.

There's also the problem that oil is being "produced" (pumped) without regard to future needs or the life span of wells. Oil-rich countries are, with few exceptions, pumping it out of the ground as quickly as they can, to feed the ever-growing appetite of nations around the world. Any country that considers holding back supplies or seizing those of a nearby country need only look at the ferocity of America's response to Iraq's seizure of Kuwait's oil fields (which, according to the CIA, have 10 percent to 20 percent of the world's known oil reserves). They'll think twice and decide not to upset the status quo.[2]

The fossil-fuel industry is driven by the common corporate objective of short-term profits, even at the expense of long-term survival. They're perfectly willing to make billions by selling off—as quickly as possible—a non-renewable resource.[3]

All this keeps oil prices low, for the time being, which discourages development of alternatives.

Another hurdle: it takes oil to make non-oil technologies

There's another problem with development of alternatives, and it's a big one. Consider, for example, solar power. Solar cells capture current sunlight, so we can use the energy immediately. But we've gotten ourselves into a bind: everything about how we produce the solar cells depends on oil:

- Solar cells are made of several rare-earth minerals, which require hundreds of tons of earth to be mined to extract a few pounds of the mineral. This mining is done with huge machines fueled by oil.
- These machines are made from materials (steel, glass, etc.) that are mined, smelted, and fabricated in oil-fired furnaces and by oil-driven machinery.
- High-temperature oil-fired furnaces are necessary to smelt and purify the rare-earth minerals once they've been mined.
- Such intense heat is also necessary to make the glass that covers the solar cells, even though the sand that is the raw material for glass is relatively plentiful and cheap. (An alternative might be to use plastic instead of glass for the covers, but plastic is made from oil.)
- And people who drive gasoline-powered cars to work and live in houses heated by oil perform the entire operation.

What happens when the oil runs out—when we no longer have stored-up ancient sunlight? Where will the solar cells come from?

We can't use today's solar cells to make more solar cells. Today's cells can barely power a small car; they certainly cannot capture enough current sunlight to create enough electricity to power a bulldozer or heat a blast-furnace or a glass factory to make more solar cells. This is a problem that environmentalists need to research and examine seriously.

We see similar problems with wind power: while there may be an inexhaustible supply of wind in some mountain passes and other parts of the country, to capture it efficiently requires high-tech turbines, built of high-quality steel and other materials that can today be fabricated only with energy derived from fossil fuels. When those parts wear out, in the absence of oil, we're left with the classic low-tech wooden windmills of Holland, extracting enough current-sunlight wind energy to pump the seawater out of a two- or three-acre field.

Currently, the production of electricity is one of our largest uses for oil. Because of this, according to the Environmental Protection Agency, electric power generation in the United States is responsible for 66 percent of all sulfur dioxide emissions, 29 percent of all nitrogen oxide releases, 21 percent of airborne mercury pollution, and 36 percent of all carbon dioxide emissions.

"Green" energy

Because of the pollution caused by burning oil to create electricity, and because oil is a non-renewable resource, there's been a growing movement in the United States recently toward producing and marketing "green electricity"—power generated by renewable sources that use current sunlight, such as wood-fired boilers, hydroelectric, solar, and wind.[4] But the demand for electricity in the United States right now is too enormous to produce from current sunlight. It's possible that we could change this, but most oil-produced electricity is so cheap it's hard for new technologies to even get started.

This makes it very difficult for companies existing in this make-money-quick-or-die economy to produce "green electricity" in a competitive fashion. With limited investment capital available because of the temporarily low oil prices, "green" power companies have sometimes had to stretch things a bit, leading to what we might call pseudo-green—power that sounds like it's green (renewable, "clean") but isn't.

The selling of pseudo-green

This became obvious at a Boston conference in May 1997, when several new "green" utilities participated in a pilot program in Massachusetts and New Hampshire to hawk their wares (electricity to your home) with hot air balloons and free spruce seedlings. The promise of the companies was that if you signed up for their power, it would be coming from a "green" source, meaning it was renewable and non-polluting. The companies advertised heavily to consumers, asking them to pay a slightly higher price for their home electricity in exchange for knowing that they were not harming the environment.

The reality of the situation turned out to be a bit different, however, as conference participants discovered when a spokesman for the Union of Concerned Scientists got up and spoke about the energy emperor's new clothes: "We think the ads in New Hampshire . . . were quite misleading," he said.

It turns out that one company was selling "hydro" power from Hydro-Quebec, a big utility that had recently flooded vast areas of Native American lands to create its watershed, against the loud protests of Native Americans and local people. Another was selling "pumped storage" hydro power, which means that massive conventional electric pumps (driven by coal-, nuclear-, and oil-produced electricity from the local power grid) were used to pump water up into a reservoir, and then it was allowed to run back out, rotating a turbine/generator to produce supposedly "green" electricity. At best, this is merely a storage system for "dirty" energy; at worst it's an outright deception of consumers. In other states, power producers have argued that they can label nuclear power as "green," since, they say, it produces no air pollution. (This overlooks the fact that it takes 18 years of continuous operation before a nuclear plant begins to generate "new" electricity—the first 18 years it's just producing an amount of energy equal to that used to mine and purify and transport its uranium fuel and to construct and maintain the plant itself.)

Such a shell game is necessary at this time because truly "green" energy is currently expensive to produce. In fact, without cheap oil to supply the industries that manufacture solar cells and turbines, it may well be impossible to produce green energy in the quantities currently required by Americans and Europeans. If we were to begin now, however, to use oil to produce systems that eliminate the need for oil, the situation could change for the better.

When fuel runs low, fighting starts

Every company in the industrialized world today, regardless of their product or service, is in some way selling goods created using repackaged oil. They're using oil to produce the electricity they're consuming, to heat the building they're occupying, to power the automobiles and buses that take their employees to and from work, and so on all the way down to the most seemingly insignificant part of their operations, such as oil being the raw material from which the synthetic-fiber carpets in their offices are made. Lacking oil, we'd be back to the level of productivity we had in 1800 when there were one-sixth as many humans on the planet and our fuel sources were vegetable oil, whale oil, coal, and wood. And when productivity plummets, resources become even scarcer.

Even a small dislocation in the availability of a primary fuel source can throw an entire nation into disarray. Many historians agree that if Hitler had had unlimited access to oil within Germany, he might well have succeeded in his plan to conquer Europe. Historical records show that Japan bombed Pearl Harbor in large part because the U.S. fleet had blockaded the waters west of Japan, cutting off Japan's supply of oil from the Indian Ocean nations. Within months, Japanese and American military strategists knew, this would have brought the Empire of the Rising Sun to its knees, although apparently American military officials underestimated how fiercely Japan would retaliate.

When oil becomes scarce in the next few decades, its price will rise, just as happened with wood in the Sumerian, Greek, and Roman empires. When the cost of the fuel source that drives everything begins to skyrocket, the small percentage of the population that controls the wealth and the armies of the world may be able to circle the wagons and protect their own interests, but the population at large will be in serious trouble. We can see this today in places like Haiti, where exploding populations have collided with limited fuel supplies and led to widespread poverty and hunger.

We in the West, being on top of the energy pyramid, will probably be the last to feel the pinch. (This presumes our armies are still intact so we can force Arab and South American countries to continue selling us their oil when supplies begin to dwindle—when the Mesopotamians, Greeks, and Romans ran low on wood, they, too, went to war. There's also the question of how we'll power the planes and tanks, if fuel oil is running low. . . .)

But even if the First World is able to use military force to guarantee access to Third World oil supplies, the dwindling worldwide fuel supply will have widespread and devastating ripple effects. Every "modern" civilization over the past seven thousand years has been crippled and then destroyed by a shortfall in their primary fuel supply. Our civilization may or may not elude the same fate. However, it's almost certain that the coming resource imbalance will strain the foundations of democracy, perhaps beyond the breaking point.

Why does it always seem to turn out this way? And what can we do about it?

PART II

YOUNGER AND OLDER CULTURES: HOW DID WE GET HERE?

In Part I we covered the evidence that we're already in serious trouble: we're all made of energy from sunlight; we're getting our energy by living off our "startup capital" (the ancient sunlight that's stored in underground "savings accounts" that are now running low); we're killing off the trees that keep our water cycle running, our topsoil in place, and our carbon dioxide under control; living species are suddenly becoming extinct in vastly higher numbers, and climate changes are leading to unprecedented unstable weather that is endangering crops and people.

How did we go, in half a century, from having an undeniably optimistic outlook to a situation where we may have just one generation left before catastrophe strikes the industrialized world, and where catastrophe has already struck the Third World, where famine stalks and disease is exploding?

There is reason for a hopeful outlook, if we make the right changes. But first we have to understand why this is. In this part we'll cover the following:

Historical perspective

- The concept of "Younger Cultures" and "Older Cultures": important distinctions between how we are today and how we used to be.
- An extraordinary cultural change called *wétiko:* after 100,000 years of living cooperatively with each other (and nature),

people began dominating and enslaving each other (and nature).

Social psychology

- An insight into the importance of "stories": how we interpret what we see around us. Our stories about life can lead us astray, or they can lead us to solutions.
- Ancient wisdom we once knew but have forgotten.
- A consideration of Darwin. Is the winner of a battle the fittest to survive the future?
- Comparing today's social and political structures with how we used to live.

To understand the pivotal importance of ancient sunlight, we first need to review the significant role that culture has played in civilizations past and present.

The Power of Our Point of View:
Older and Younger Cultures

 A chain is no stronger than its weakest link, and life is after all a chain.

—WILLIAM JAMES (1842–1910)

Because of human actions—and inaction—our planet appears to be on a collision course with disaster.

We long ago passed a human population number that could be sustained without intensive use of gasoline and oil, so we're burning up a 300-million-year-old fossilized-plant resource in order to feed the six billion humans currently riding spaceship Earth. Many more may starve, even more than are starving today. And almost nothing is being done by governments to offset this very real possibility.

But what can we do? We recycle, eat vegetarian foods, drive gas-efficient cars, and feel as though we're doing something useful, but it remains a fact that a panhandling Bowery wino in New York City has access to greater wealth in a month than most citizens of the world's population will ever see in one year. And even that "poverty-level" rate of resource consumption is something the planet cannot sustain without our burning up carbon fuel sources that will be exhausted within a generation or two.

Most people probably believe there is nothing they can do to help lessen the burden. But they are mistaken: there are indeed powerful, meaningful things we can do.

Perhaps it's too late (by at least four decades, according to many experts) to avoid *all* of the damage: the death of billions of humans and further extensive destruction of much of the planet's environment through war, natural resource exploitation, and industrial pollution. Earth is now afire with war; famines are erupting as you read these words, and overpopulation has soared to the point where, in many of the bigger cities of the Third World, street children are hunted with high-powered rifles by "hunting clubs" of middle- and upper-class young men and off-duty police. Some speculate that we are witnessing the last days of the American/European empire, much as the Romans watched their empire crumble 1,600 years ago.

But it is also true that we can plant the seeds of a positive and hopeful world where future generations, our children and theirs, will live—the beginning of the next civilization, the post-oil era.

We hold their future in our hands.

There's power in how we think about things

When you walk or drive down a city street, what you are seeing all around you are manifestations of thoughts. Every building began as an idea in somebody's mind. Someone acquired the land. Someone designed the house. Someone had the idea to organize people together to build the house, either to make money or to live in it. The trees you see were planted for shade in the yard, on the sidewalk, along the street. The pavement that we accept as a "natural" part of our landscape was conceptualized, designed, engineered, installed, and is maintained via thought.

Thoughts create our physical reality, and they also create our larger reality. During ancient times, when there was an electrical storm, people perceived the thunder and lightning as the voice of a powerful deity. If somebody got hit by lightning, that proved to others that the person had committed some crime or displeased the deity. When the thunder rumbled loudly nearby, people knelt to the ground and cried out their prayers. They knew, when they saw the awesome streaks of light across the sky, that they were seeing the finger of their god writing messages or expressing an opinion. Today thunder and lightning are seen and heard as the discharge of electrical energy between ions in the air and the oppositely charged ground.

If somebody is struck by lightning, it is either due to their own stupidity (standing on the golf course with a club in the air) or just bad luck. If the storm is severe, we take cover out of fear of a dangerous natural phenomenon, rather than a wrathful god. The same event creates a completely different feeling, thought, and behavior in the people who observe it today. The point is, the experience of reality is different, and what makes it different is thought.

A few years ago, I was invited to give a speech at a conference sponsored by the Hebrew University in Jerusalem. After the presentation, my wife, Louise, and I went for a walk through the Old City, down through the Arab Quarter where most of the tourist shops are. It was a Friday, the Moslem Sabbath, but not being Islamic, we doubted this would affect our sightseeing and shopping. It being a very hot May day, Louise was wearing a comfortable pair of walking shorts. As we moved through the streets, one shopkeeper came out of his store and started shouting at Louise, calling her a "Western pig," a "whore," and a "blasphemer." "Don't you know it's a holy day, you bitch?" he screamed. "You have no right to show your legs!"

I mention this culture clash because the shopkeeper's reality was that a woman (who, in Islam, has a different social status than women in most Judeo-Christian or Western cultures) was flagrantly breaking the law. Louise's reality held that it was a hot day in a tourist center and she was dressed comfortably in conservative shorts, according to Western standards, and being harassed for it. My reality told me that a man was displaying bad manners and disrespect for my culture and religion, for women in general, and for another human being in particular, by shouting instead of quietly coming up to us and presenting his case.

We were all correct.

And so now all of humanity is presented with a dizzying set of conflicting realities. What we choose to do about them will determine our future as a species. Consider these various ideas different people might have about life:

- "We need electricity to be comfortable and maintain our way of life," or "Producing electricity is pumping billions of tons of carbon dioxide into the atmosphere, leading to global warming and extremely destructive weather patterns."

• "Being able to drive where and when we want to at a low cost is freedom," *or* "Americans' driving habits are feeding the destruction of the planet."

• "All of nature is here to serve the needs of humankind," *or* "Humans are no more or less important to the planet than any other life-form."

These ideas are grounded in the *stories*—the myths of our culture, our paradigms, our beliefs—that form the core of what we tell ourselves is "reality." Stories, in this context, are anything we add to our original experience that alters what we think is going on, or changes how we think about things.

Since so much of what we call reality is subjective, there are few "right" or "wrong" stories; instead there are "useful" and "not useful" stories, depending on what culture you belong to, and depending on your status in your culture. Depending on your relationship to the natural world and your vision of the future.

Increasingly, the stories we've been telling ourselves for centuries are now moving from the "useful" to the "not useful" category. An example of such a story is the biblical order to have as many children as possible. In the days of Noah and Abraham, the tribe with the largest number of young men to create an army was usually the tribe that survived. "Be fruitful and multiply" was a formula for cultural survival, even though in nearly all cases it then led to "and when you run out of resources and living space, kill off your neighbor and take theirs."

We've rationalized this over the years by saying that this conquering and dominating lifestyle has brought us so many "good things": television, visiting the moon, modern appliances, the eradication of many diseases. I remember in high school a recruiter for the U.S. Army came in and gave a pitch for the armed forces to our tenth-grade class. "Most of the really important advances in our civilization, from the development of rockets to the discovery of antibiotics, were caused by the necessities of war," he said, providing another feel-good rationalization for the periodic mass murder of humans. War is good: it leads to progress and lifestyle upgrades.

Back when the planet had only a few million people on it, a person could probably have built a case for the value of huge families, growing populations, and the conquest of nearby (or distant) lands. It might have been of questionable morality, but it could have been defended by the norms of a culture that had survival and growth as its primary goals.

Now, however, such stories imperil the very culture from which they're derived. Over a billion Catholics and over a billion Muslims are both still acting out this story, and the results for the planet are increasingly clear in Catholic South and Central America and the Muslim areas of the Middle East and southern Asia.

The ancient Greeks changed the world and established the foundations of Western Civilization with the idea of integrating democracy and slave ownership. In fact, every time a culture has been transformed, since or before then and for better or worse, it's been because of an idea, an insight, a new understanding of how things are, and of what is possible. Ideas preceded every revolution, every war, every transformation, and every invention.

So, the good news is that if we redefine our cultural norms, retell the stories that make up the reality we follow, then humanity's behaviors will change to conform to the new stories.

First, though, we must understand both our current and past stories, in order to create future ones that can work.

Younger Culture Drugs of Control

 It is not heroin or cocaine that makes one an addict, it is the need to escape from a harsh reality. There are more television addicts, more baseball and football addicts, more movie addicts, and certainly more alcohol addicts in this country than there are narcotics addicts.

—SHIRLEY CHISHOLM (b. 1924)

Politicians and writers often refer to our current era as the Information Age. The average person alive today, they say, knows more than anybody at any time in the past. Through the Internet, encyclopedias on CDs and DVDs, and with access to 700-channel television, the collective knowledge of the planet is available instantly to even the most ordinary of citizens, they say. It's a wonderful thing, and we're spectacularly well informed.

But is this really so?

If we are so well informed, why is it that when you ask most Americans simple questions about the history of the world, you get a blank look? How many of our children have read even one of Shakespeare's plays all the way through? How many people know with any depth beyond the 15-second sound bites served up in the evening TV news the genesis and significance of the wars in, for example, Bosnia or the Congo? Or that the United States government is still stealing Indian lands in Nevada, Minnesota, Wyoming, Arizona, New Mexico, Alaska, and a dozen other states?

Yes, the Internet is an extraordinary source of information. (I use it myself for research, and have a website for my books.) But the vast majority of profitable, well-used Internet sites are the ones selling sex or pornographic photos. The most common "search" word reported by *all* the "search engine" companies is "sex," with other words describing nudity or various sexual acts occupying the rest of the top of the list. Next on the list of most-used are the sports and entertainment "channels," complete with interactive soap operas and clips from the latest movies and TV shows. The simple fact is that the Internet has not had much impact on how well informed Americans are.

What about television, then? Occasionally in speeches when I say that years ago we disconnected the TVs in our house—not because of the content but because the medium itself can create progressively shorter and less functional attention spans in children—people in the audience will go to great lengths to tell me how much they've learned about, for example, African wildlife, by watching PBS specials. I don't disagree that there is some interesting and informative programming on TV. But it's only in programs, and there's far more information in books—and recent studies show that only a tiny percentage of Americans have read a book in the previous month. When former FCC commissioner Newton Minnow described TV as a "vast wasteland," he was understanding it. In the rush for advertising dollars, which are priced based on viewership numbers, TV shows are becoming more and more entertainment-rich and information-poor (including the "news"). And even the "information" on TV is too often there to serve a particular corporate interest.[1]

Balanced and fair lies?

Media consolidation has led to a virtual absence of dissenting voices on radio, TV, or in newspapers.

For example, Congressman Bernie Sanders (I-Vermont) points out that in 1993 when NAFTA first came up for a vote in the U.S. Congress, even though polls clearly showed that more than half of all Americans opposed it, of the hundreds of corporate- and chain-owned newspapers in cities across America, only *two* editorialized against it—the rest all ran both coverage and editorials in support.

The U.S.-led invasion of Iraq in 2003 saw a repeat performance of corporate-owned newspapers molding rather than reflecting public opinion, when over 60 percent of Americans were opposed to unilateral intervention without U.N. sanction, yet *none* of the chain-owned newspapers across America editorialized against the invasion.

Some suggested this was because war is good for the news business—ratings go up and advertising revenues follow. Others pointed out that some multinational corporations that own broadcast networks and media outlets are also big players in the defense industry, and their weapons business is often their most profitable division. Some suggested that news outlets were successfully cowed by bully tactics engaged in by the White House, denying press credentials or refusing to answer questions from "hostile" reporters such as Helen Thomas, while granting high levels of access and news scoops to "friendly" reporters. The most cynical of media observers suggested that what was really going on was a down payment to the Bush administration in exchange for its modifying media-ownership rules, due to be decided in June 2003 by the FCC, which were preventing the final and total takeover of American media by a handful of the world's largest corporations.

Information deficits

We may be living in an "information age" with "information overload," in some sense. But when it comes to what actually gets into people's heads, we're living in an age of "knowledge scarcity." People no longer know information that's vital to sustain life such as how to grow their own food; how to find drinkable water; what's in their food; how to build a fire and keep warm; how to survive in the natural environment; how to read the sky; when the growing seasons begin and end; what plants in the forest and fields are edible; how to track, kill, dress, eat, and store game; how to farm without (or even with) chemicals and tractors; how to treat broken bones and other common medical emergencies; or how to deliver a baby.

Because of this "information deficit," we are out of touch with reality and are also standing on a dangerous shelf of oil-dependent, corporate-induced information starvation. In the 1930s during the

Great Depression, far more people lived in rural areas than in the cities. The information about how to grow and preserve food, how to survive during difficult times, and how to make do with less was general knowledge. Today we know the names of the latest movie stars and how much their movies grossed, or what level the Dow Jones Industrial Average is at, but few of us could survive two months if the supermarkets suddenly closed. In addition, according to the Barbara Bush Foundation for Family Literacy, fully 27 percent of all American adults are "functionally illiterate" . . . although fewer than 1 percent of American homeowners lack a television set.

This works to the tremendous advantage of anyone who'd benefit from our being dependent on their systems, information, fuel, and food. We've become easy to manage and easy to control. We'll vote for whomever has the best 10-second sound bite on the evening news, or the most powerful and expensive advertising.

Companies that sell or produce toxic or carcinogenic chemicals as waste by-products are able to spin or suppress news stories so effectively that most citizens of Vermont, for example, don't know that over 53,000 pounds of a chemical banned in Germany, Italy, the Netherlands, Sweden, Austria, and other countries is sprayed on their feed corn every year. This chemical has been linked in scientific studies to breast cancer, leukemia, lymphoma, birth defects, and reproductive tumors. On the same day that story was featured in an advertisement by the Vermont activist group Food & Water, a news story in a different part of the same day's newspaper mentioned that scientists had received numerous reports of malformed frogs on Vermont's Lake Champlain.[2] The birth defects included "missing or deformed limbs," "frogs with eyes staring from their backs," and "suction cup fingers growing from their sides." Yet even with all the activist groups who are trying to get the word out, it may seem that we might as well be asleep, for all the good it does us.

We're not just asleep; we're intoxicated

As a teenager growing up in the sixties in college towns and San Francisco, I made the acquaintance of several heroin addicts. By and large they were nice people—not the stereotypes we see on TV and in literature, but relatively normal middle-class kids who got in over their heads with a drug that was stronger than they'd ever expected or believed. Later, as I grew through my twenties and thirties, I met my share of alcohol addicts. Similarly, most were good people at heart, but had found themselves in the grip of a drug that consumed their lives. And I've known many tobacco addicts over the years, most similarly well intentioned, who always thought they could one day just say no, and then discovered it was unbelievably difficult.

One of the things that I noticed about these addicts was that keeping the supply of the drug flowing into them had become the most important thing in their lives. It was at the core of their existence. They'd wake up in the morning and their first thought would be filled with how to get that day's supply of their particular drug. The day was drenched with the drug, and eventually they'd go to sleep with their drug.

Another thing you notice about addicts is that they will sacrifice things that they might otherwise consider important for their drug. They may have great plans for career, education, or relationships, but somehow those things end up subordinated to the enjoyment of their drug. Long after the drug has stopped producing a "high," but is just keeping them from flipping over into painful withdrawal, they're still spending hours every day immersed in their drug.

From the point of view of those running our culture, this is considered a good thing: Younger Culture governments have traditionally regarded it as desirable to get people addicted.

For instance, consider that the U.S. government continues to give millions of dollars in subsidies (a nice euphemism for "gifts" or "corporate welfare") to tobacco producers. In the more distant past, 30 years after losing the American Revolutionary War, the British fought a war with China to protect their "right" to sell opium to the more than 12 million addicts they'd created in China. They won the war, and took Hong Kong as part of their booty, and the empire made billions on opium trade and opium taxes. Many historians believe that the British were successful in winning the Opium War in large part because so many members of the Chinese royalty and bureaucracy were themselves addicted to opium. This both reduced their effectiveness as military opponents and reduced their enthusiasm for making the British—and their opium—go away.

In dominator Younger Cultures, the first goal of the culture itself, as acted out most often by the cultural institutions of government and religion, is to render the citizenry non-resistant. Earlier we saw that what typically happens to peoples who won't "adjust" is that they're exterminated. Many Native peoples have shared this fate; the result is that the only conquered peoples who survive tend to be docile. (If it sounds like conquerors treat the conquered like animals to be domesticated, you're getting my point precisely.) As every heroin dealer, tobacco salesman, and liquor store owner knows, if you have people who depend on a daily dose of your product for a sense of well-being, you have people who are not going to give you much trouble. (They may cause problems for others, but generally the dealers are left alone.)

Similarly, our technological culture has found a technological drug to maintain docility.

One measure of a drug's addictive potential is what percentage of people can take it up or put it down at will and with ease. This behavior is called chipping a drug—occasionally using it, but also walking away from it without pain or withdrawal for months or years at a time. Research reported in *Science News* found that while large percentages of people could chip marijuana, and medium percentages of people could chip alcohol, cocaine, and even heroin, very, very few people (less than 5 percent) could chip tobacco. But imagine a "drug" that fewer than even 5 percent of Americans could walk away from for a month at a time without discomfort. Such a drug, by the definitions of addiction, would be the most powerfully addictive drug ever developed.

In addition to discouraging chipping behavior, this drug would also have to stabilize people's moods. It would put them into such a mental state that they could leave behind the boredom or pain or ennui of daily life. It would alter their brainwaves, alter their neurochemistry, and constantly reassure them that their addiction to it was not, in fact, an addiction but merely a preference. Like the alcoholic who claims to only be a social drinker, the user of this drug would publicly proclaim the ability to do without it . . . but in reality would not even *consider* having it be completely absent from his home or life for days, weeks, or years.

Such a "drug" exists.

Far more seductive than opium, infinitely more effective at shaping behavior and expectations than alcohol, and used for more minutes every day than tobacco, our culture's most pervasive and most insidious "drugging agent" is *television.* Many drugs, after all, are essentially a distilled concentrate of a natural substance. Penicillin is extracted from mold; opium, from poppies. Similarly, television is a distilled extract—super-concentrated, like the most powerful drugs we have—of "real" life.

People set aside large portions of their lives to watch a flickering box—hours every day. They rely on that box for the majority of their information about how the world is, how their politicians are behaving, and what reality is, even though the contents of the box are controlled by a handful of corporations, many of which are also in the weapons and tobacco and alcohol business. Our citizens wake up to this drug, consume it whenever possible during the day, and go to sleep with it. Many even take it with their meals.

Most people's major life regrets are not about the things they've done, but about the things they've not done, the goals they never reached, the type of lover or friend or parent they wished they'd been but know they failed to be. Yet our culture encourages us to sit in front of a flickering box for dozens (at least) of hours a week, hundreds to thousands of hours a year, and thereby watch, as if from a distance, the time of our lives flow through our hands like dry sand.

The sickness of "living in boxes"

Psychologists agree that being separate from others is generally harmful to our mental health and well-being. To be well, we must connect with others.

Louise and I live with a cat named Flicker, a beautiful neutered black female with a thick gray mane that makes her look like a miniature lion. Flicker is nuts. The person we got her from told us that Flicker is quite certain that every human in the world is out to kill her and, we found, that appears to be true. A "scaredy-cat," she is paranoid, in the clinical sense.

Yesterday, I came across Flicker in the hall, on my way to the living room via the kitchen. She looked at me in bug-eyed fright, spun around, and ran toward the kitchen. I was heading in the same direction, so kept walking: now she was *certain* that I was coming to get her. In the kitchen she paused for a moment, but I kept coming, as the way to the living room is through the kitchen. She glanced around with a panicked look, then ran toward the living room: I was still behind her. I tried purring at her, making soft sounds, and calling her, but nothing works with this cat: she knew that I was coming to hurt her. In the living room I encountered her again, which sent her flying up into the air and then out to the safety of another hall that leads to the front door. Flicker's world is a hostile place filled with malevolent giants. In the few months we've had her, we've managed to get close to her from time to time, but there is always that wildness in her, that latent certainty that she can trust only herself for her own safety.

I was a guest on a nationally syndicated radio show a few weeks ago, talking about some of the issues in this book, and a man called in from someplace in Kansas.

"Do you mean to say," he said, "that plants and animals have a *right* to life on this planet?"

"Yes," I said. "That's exactly what I mean to say."

"You know that that's the position of the 'deep environmentalists,' don't you?" he said. "The radical tree-huggers?"

"Yeah, I've heard that," I said. "What's your position?"

"That we have to assign a value to things, using science and economics. Some forests are worth keeping and others are not. Some species can survive along with us, like cows and dogs and deer, and others can't, and so we shouldn't worry about them."

"So where do you draw the line?" I asked. "How do you know which species we should keep and which we should wipe out to make more room for the ones we like or to make more room for more people?"

"Keep the ones that are useful!" he said, as if the answer were obvious. "Who needs a spotted owl or a snail darter, for God's sake? We need jobs, economic security, clean streets, and safe cities. Those are the important things."

I pointed out to him that even if his assumption (that the world is only here for humans) was true, such massive tinkering as wiping out hundreds of thousands of species and altering the chemistry of the atmosphere might still create unintended results that would end up making the planet inhospitable to our "master species." And, in fact, there's plenty of evidence that that's exactly what's happening, documented in this book and many others.

If we were to set aside the assumption of our supremacy and instead adopt the Older Culture view of all things having value and a sacred right to live on this planet, then the odds of our unwittingly taking planet-scorching actions plummet.

Like Flicker, the caller to this radio show sees only one world. That world is a place populated by bright and colorful and "real" human beings, and every other living thing has a dimmer presence. Every "thing" is here to serve us, and we are given the knowledge and power over what shall live and what shall die. If it is to our advantage to strip the world naked, down to a single species of tree and grain and vegetable and fish, then so be it. We have decided that it is right, because we see and understand the world as it really is. And for those who don't believe it's possible, we have the words of several of our gods, reported by humans who are incapable of error, to prove it.

This is the logic of the mentally unhealthy or ill.

Just as Flicker is certain she has the world figured out, and that my walking from the bedroom to the living room—regardless of what I may think my motives are—is proof positive of the malevolent intent of all humans, the caller is certain that everything he sees in the world was put here for him, and if I assert it has its own right to existence, then I am conspiring to take it away from him. Paranoids construct a detailed and well-organized world where everything makes sense and is self-reinforcing. That man on the corner who is looking at you is the CIA spy who put the transmitter in your brain.

He looks away because he doesn't want you to know that he is the spy. He glanced at you not because you were staring at him but because he is wondering if you have figured out that he is responsible for the transmitter. He gets on the bus not to go to work but to follow you. And on and on.

Similarly, whatever our worldview, we collect evidence to support it. Flicker believes people are chasing her, and sees signs of it everywhere. So if you believe everything is a resource that we can use to our advantage, you'll see signs of that everywhere, too.

Sigmund Freud, the father of psychoanalysis and the man to whom many today look for definitions of what is "sick" and what is "healthy" mentally, made some interesting observations along these same lines in the years before he died. He pointed to his belief that what our civilization refers to as a "healthy ego" is, in fact, "a shrunken residue" of what we had experienced early in life when the ego experienced a "much more inclusive" and "intimate bond" with the world around it.[3] Many psychologists say that one result of this "shrinking process" is that the third most common cause of death for Americans between 15 and 27 years of age, according to the National Institutes of Mental Health, is suicide.

This shrinking into separateness, this breaking of the intimate bond with the world around us, this separating ourselves into isolated "boxes," was largely unknown for the first 100,000 years or more of human history. It is still largely unknown by tribal people around the world, which, among those who have little contact with Younger Culture people, have a suicide rate so low as to be almost unmeasurable.

University of California at Hayward professor Theodore Roszak uses the word "ecopsychology" to define the study of the relationship between humans and the natural environment. In his books *The Voice of the Earth* and *Ecopsychology,* Roszak eloquently shows how the physical, mental, and spiritual disconnection of modern people may be responsible for entire realms of personal and cultural mental illnesses, and how reconnecting with nature can be a powerful therapeutic process both for the individual and for society.[4]

But this disconnection from nature has been at the core of "civilized" human experience since the formation of the first such "civilization" seven millennia ago. It was celebrated by Aristotle in his writings on how the universe and natural world were merely collections of simple particles (atoms) that humans could manipulate once they understood them, and refined by Descartes, who argued that the entire universe was a giant machine, and this machine-like nature echoed all the way down to the smallest level. If we could just figure out where the levers and switches were, we could always figure out a way to control the machine. We withdrew from the natural world and created an artificial world around us, in our cities and towns, which is quite alien from that in which we first evolved. We even asserted that animals were just biological machines, incapable of feelings or emotions. As time went by, we decided for ourselves that various things were right and wrong with the rest of the planet, and set about organizing things "out there" to comply with our needs "in here."

We placed our planet at the center of the universe and ourselves at the top of the hierarchy of our world. Our Younger Culture religions and philosophers proclaimed, both explicitly and implicitly, that all of creation is made only for man. Galileo even went so far as to propose that if humans were not present to observe the world, it would cease to exist. When it was finally accepted that our planet wasn't the pivot-point of all creation, we simply shifted our language to accommodate a fundamentally unchanged worldview: it is now the assumption of almost every "religious" citizen of any "civilized society" that we are at the *spiritual* center of the universe.

From this story, this view of the world—that our man-made cities are civilized and the natural world is wild and people who live in it are primitive or uncivilized or savages—we have developed a psychology that acknowledges and praises only itself and its own culture and has lost contact with the real physical world and its extraordinary powers and mysteries.

When the early European/American settlers fanned out across the prairies and killed every buffalo they could find, the Native

Americans watched in shock and horror at what they considered a senseless act of insanity. How could the settlers take the life of the plains? How could they parcel up the flesh of Mother Earth? How could they be so crazy as to cut down every tree in sight? The settlers looked at the Indians and thought they were crazy to not take and eat all the buffalo they could. How could they have sat on this valuable resource for 10,000 years and not have used it? They had to be savages, uncivilized half-humans who didn't have the good sense to know how to use nature's bounty for the good of the human race.

For a while, this worked for the conquering "Americans." Just as Gilgamesh could cut down the cedars of Lebanon, just as the Greeks could destroy their own forests, just as Americans could strip half their topsoil from the land, the rapid consumption of "out there" to satisfy the needs of us "in here" worked for more than a few generations.

No more, as we're seeing in the "early warning system" of the Third World. In our inner cities where people are afraid to drive with their doors unlocked or windows down, on our farms where dioxin or PCB-laced waste is spread across food plants as fertilizer, in our hospitals where the primary waste from the manufacture of nuclear weapons (yttrium) is being promoted as an experimental "cure" for cancers (which are caused in large part by the air and food and drugs of our civilization)—in all these places we see that this world we have created can work only for a very few. It is the nature of hierarchical, dominator systems to end up that way.

Older Cultures are older because they have survived for tens of thousands of years. In comparison, Younger Cultures are still an experiment, and every time one has been attempted (Sumeria, Rome, Greece), however great its grandeur, it has self-destructed, while tribes survive thousands of years.

Younger Cultures are built on a foundation that is psychologically and spiritually ill: Freud's "shrunken residue" of the true and historic beauty of human life lived in intimate connection with the natural world. Increasingly, we live in isolation, in "boxes"—and suffer for it.

What it's like to be in touch with the world again

It's possible to climb out of the box and get back in touch with the world.

Over the past 25 years, I have taken several classes in wild edible or medicinal plants. Usually they involved one or more trips into the forest and fields in search of the plants being studied. One of our teachers carried with her a small bottle of yellow cornmeal. She said, "When I uproot a plant or cut off a leaf, I put some cornmeal on the earth as my way of acknowledging the spirit of the plants, and giving an offering back to them for their giving some of themselves to us."

In *The Origin of Consciousness in the Breakdown of the Bicameral Mind,* Columbia University psychology professor Julian Jaynes puts forth the concept that in prehistoric times (more than 7,000 to 10,000 years ago) people actually heard the voices of the gods.[5] When they looked out into the natural world, they saw fairies, sprites, spirits, and other entities. This was because, Jaynes posits, the two hemispheres of the brain were more fully connected, so that the auditory regions of the left hemisphere were directly connected to the hallucinatory regions of the right hemisphere (Wernicke's and Broca's areas) that, in modern people, are normally active only during dreaming or in schizophrenics. Because of this direct connection, Jaynes suggests what we now call hallucinations probably were a common part of the everyday experience of ancient peoples.

It was the rise of the Mesopotamian city-state empire, Jaynes suggests, and its use of written language that was largely responsible for the breakdown of this connection between the two hemispheres of the brain, causing all of us except the occasional mystic or schizophrenic to lose contact with much of the right hemisphere during our normal waking consciousness. Jaynes's arguments are persuasive, particularly where they draw on historical record and contemporary neurology. If his perspective is accurate, then we would expect that people living today the same way all humans did 10,000 years ago would live in a world alive with spirits, energies, and voices. When people are removed from that world and "civilized" by learning to read and write, they would quickly (in as little as one generation, perhaps within an individual's life span) lose contact with that other world.

Another view is advanced by Terence McKenna in *Food of the Gods*.[6] McKenna believes that the reconnection of the bicameral mind in cultures ancient and modern was and is brought about by the ingestion of certain plant substances. Hallucinatory plants are used by numerous cultures to open the doors to the world of the gods, McKenna points out. He even goes so far as to suggest that the rigidity, pain, and sterility of modern life is largely the result of our having lost access to those worlds because of the regulation and control of these substances that once grew throughout humanity's habitat. McKenna proposes that the use of these plants helped catalyze the birth of human consciousness in early primates. This, in turn, spurred the development of the thinking and mystical brain/mind, and gave the human species the mental power to set about replacing the plants with its own ways of controlling the mystical or divine experience, principally through the force of law promoted by organized religions.

Both Jaynes and McKenna contribute significantly to our understanding of the history of consciousness. McKenna has lived among and studied tribal peoples who today use these plants to meet and talk with the spirits of their world, and Jaynes has extensively studied the writings of past civilizations and people who said they heard the voices of their gods within their own heads. Regardless of the technique or method, there is a consensus among them and others that ancient and "modern primitive" people share the ability to see, feel, and hear something that we in modern Western society generally do not.

When a Shoshone looked about for food, he listened to what the land told him, the voices of the plants and animals and the Earth itself. They showed and told him where his day's meal would be found, and also what types of ceremony would be appropriate to thank the world for this gift.

Contrast this with how European kings lived in the Middle Ages, and how the dominator mind-set of that era has led us into an ironically unaware pseudo-"information age" and, perhaps unwittingly, into what author Daniel Quinn and the Australian Aborigines call "the great forgetting."

* * *

Our minds and our cultures created our situation. There's great insight in understanding this, and power in realizing how much of a role we can play in redefining the future of the planet for ourselves and our children.

Younger Culture Stories About How Things Are

 Think of the earth as a living organism that is being attacked by billions of bacteria whose numbers double every forty years. Either the host dies, or the [parasite] dies, or both die.

—GORE VIDAL, "Gods and Greens," *Observer* (London, August 27, 1989)

I remember 1960. John F. Kennedy had just been elected president of the United States, and promised to turn our nation away from the policies of exploitation and segregation that had been institutionalized during previous administrations. Kennedy challenged us to think of our children's future before our own, to create change that could be lasting, to build a new world that could sustain itself and yet protect its valuable natural resources.

According to the U.S. Census Bureau, in 1960 there were 3,038,930,391 humans on Earth. That year saw the addition of 40,622,370 people—each one requiring three meals a day, several gallons of water a day for drinking and bathing, and a place to live. While the governments of the world struggled to keep up with and meet the needs of these 40-plus million new Earth citizens, in 1961 another 56,007,855 more people were born than died. And as we struggled to find a place for them, in 1962 we increased the world's population by another 69,393,370. The year 1963 saw another 70,987,231 humans competing for the food, water, shelter, and heat of our planet.

From the three years between the time John Kennedy was sworn into office until the day he was shot in November 1963, more people than the entire population of the United States were added to the world. It was the upsweep of the explosion that would soon see almost every habitable territory on this small blue ball floating in space filled with human flesh.

It's easy to wax nostalgic about "the good old days": almost every generation has done it for as long as we've recorded history. But the fact of the matter is that in 1960 you could travel just about anywhere on the planet with some assurance of safety: you could hitch-hike across North, Central, and South America and survive the experience (as many did), and famine was a localized and unusual phenomenon.

Today none of that is true. And today there are about twice as many people on Earth as in 1960.

Now, if you can imagine it, we're about to repeat that experience. In another thirty years, we'll add as many people to the planet as were alive in 1960. And as the resource pot gets smaller, those fighting to keep their share of it become increasingly more desperate. More violent. More deadly.

Today's "Younger Culture" view

Our culture is young: after the hundreds of thousands of years of human existence, our culture carries an idea, a story, which came to us from just the past seven thousand years. It's our collective story of how things came into being and how they'll continue or end. Modern consciousness has been formed by a set of very specific myths, beliefs, and paradigms (which I call *stories*), and it's these stories that have brought us affluence and comfort, yet, paradoxically, are speeding our culture's demise. They can be summarized as follows:

• **We are not an integral part of the world; we are separate from it.** The Earth (and all of the plant and animal life on it) is something different from us. We call that different stuff "nature" and "wilderness"; we call ourselves "mankind," "humankind," and "civilization." We are very clear in our vision of the difference: we are separate from it, superior to it, and a law unto ourselves. When we want something, it's there for us to take, and we don't have to answer to anyone else.

• **It is our destiny to subdue and rule the rest of creation.** From the Bible's command to establish "dominion" over the Earth and its inhabitants, to the American government's acted-out doctrine of Manifest Destiny, to our science-fiction stories about colonizing space, we tell ourselves many stories which express that we deserve to be the designated rulers of everything we can see, from the seas to the moon and beyond. Some people try to soften this by saying that when Man was given dominion of the Earth, it meant he was given responsibility for taking care of it, but few people in our culture behave as if they believe this.

Younger Cultures see themselves as dominators and conquerors. They don't just live in their own area, support themselves, and defend themselves against invasions; they seek out opponents (animal or human) and capture, enslave, or eradicate them. Their agriculture seeks to pump as much nutrition out of the soil as possible, even if they leave the soil dead. They keep things away from each other, and they have police and armies to help those with wealth keep it intact.

These ideas are reflected in the writings of our culture's foundational and seminal thinkers. Aristotle defined the classical Greek view in his essay entitled "Politics": "Plants exist for the sake of animals, and brute beasts for the sake of man—domestic animals for his use and food, and wild ones (most, at any rate) for food and other accessories such as tools and clothing."

The Roman perspective was summarized well by Cicero, who wrote: "We are the absolute masters of what the earth produces. The mountains and plains are for our enjoyment. The rivers belong to us. We sow the seeds and plant the trees. We fertilize the soil. We stop, direct, and turn the rivers; in short, by our hands and various operations in this world we make the world as if it were a different nature."

By the 1600s, Francis Bacon made it explicit in *Novum Organum* when he wrote: "I am come in very truth leading to you

Nature with all her children to bind her to your service and make her your slave."

In the nineteenth century, Karl Marx wrote that the goal of socialism is to be "rationally regulating their (humanity's) material interchange with nature and bringing it under their common control." Engels referred to humans as "the true masters of nature."

At first, we applied these beliefs only to nature. Instead of simply competing with other animals for our food supply, we began to rule them. Since we ruled them, we could go beyond simply competing with them to the point of working to totally destroy them . . . which we did to every species that competed with us or interfered with our getting food or food-producing land. From wolves to insects to weeds, we developed new and better ways to wipe out our competitors.

From this point of view, it's not a great leap to apply the concept to other humans. The logical extension of the idea that humans have the right to rule all creation is to believe that some humans are "more rightful rulers" than others. And since we'd already decided that it was not just acceptable but *good* to destroy competitors, we developed better and better ways of destroying other humans, which have reached their most sophisticated expression in our modern engines of war.

Another way our culture reinforces our worldview is found in the structure of our language. When Dorothy Lee lived with the Wintu tribe in northern California in the 1950s and learned their language, she was intrigued to discover that their verb-forms were largely lacking in terms of ownership or coercion.[1] Instead of saying, "my sister," one would say, "I am sistered by," or, "I live with a sister."

Most Younger Cultures are quite different. Look at the various beliefs around the world about the relative roles of men and women, for example. In some countries, particularly those under Islamic law, women are not allowed out alone, must cover their bodies with clothing, cannot vote or run for office or hold power in any way outside the household, and are essentially the property of men.

The murder of women who have "disobeyed" their fathers is not uncommon in Muslim and Hindu countries, as was shown in August

1997, when Marzouk Ahmed Abdel-Rahim hunted down his 25-year-old daughter, Nora, and beheaded her, then carried her severed head back to his village near Cairo, Egypt, and pronounced to the assembled villagers, "Now the family has regained its honor." Her crime: she had eloped. The next day, in a village 40 miles north of Cairo, another man set his 19-year-old daughter on fire by pouring gasoline on her and throwing a match at her in front of the community. "Nora was not an isolated victim," the Associated Press reported on August 19, noting that in many such cases the child's choice of spouse doesn't have enough money to pay the family a dowry, and, according to Egyptian writer Nawal Saadawi, in these cultures "Marriage is viewed as a business transaction, and the goods are the women—to be bought and sold as the father sees fit."

Wétiko: gaining by consuming others' lives

Ten thousand years ago, this was a new and radical idea, that humans could "own" and "dominate" nature or each other. We see little evidence of it even today among the surviving "Stone Age" and/or tribal people who are living the way humans lived 50,000 years ago. In the 1990s, for example, the Dani people of Indonesia faced a crisis because their culture taught that it was wrong for humans to own land (or almost anything—they've lived tribally and communally for at least 12,000 years). Because they were unwilling to assert ownership claims to the land they'd been living on since before recorded history, because they considered such an act immoral and wrong, outsiders came in and filed legal claims for their "unclaimed" land and began to cut down the rainforest in which the Dani live.

Similarly, in 1997 I spent three days with several hundred Native Americans from the Oneida, Choctaw, Cherokee, Ho-Chunk, Navajo, Poma, Ojibwe, Otoe, Cree, Lac Couste Oneilles, Mescalero Apache, Tohono O'odham, and Cheyenne River Sioux tribes (among others). Over and over again, I heard people talk about the need to respect "our Mother" the Earth, to experience a connectedness with nature and the connectedness with Spirit that comes through being connected with nature. These are people from what I will later in this book refer to as "Older Cultures," who even to this day see the Earth as a sacred place and respect the rule of nature that one must live in harmony with one's environment, and not dominate and destroy it.

So after 100,000 or more years of humans living largely in harmony with Nature, there were these cultural eruptions: in a few parts of the world some tribes began to dominate nature, to intervene in restructuring their surroundings, to begin converting huge amounts of land from natural forest into grazing or farming land, thus producing even more human food.

While this new Younger Culture behavior left less food for other animals, and decreased the overall number of living species because we cleared "inedible" plants (mostly trees, thus also wiping out their species-rich ecosystems) to replace them with single-species human- or livestock-edible plants, it made more human food available.

The notion that we're "of course" destined to dominate whatever we can had far-reaching implications. This "story about how things are" literally changed the world, often in very brutal ways. It was the basis for the philosophy used to rationalize the Imperial Conquest of most of the world by Britain. It was the basis of Manifest Destiny in America, in which Congress decided that their idea of God wanted us to take over the continent, so we killed anyone who got in our way, resulting in the murder of tens of millions of Native Americans.

And it's not just Americans; the same has been done to indigenous peoples worldwide. Slavery, apartheid, and the entire concept of Darwinian economics have been used to justify continued suffering by masses of human beings.

There are rulers, and there are those who are ruled, we noticed. This is natural law, we believed. It must be that this is the way things were organized from the beginning, and so we are only acting out the way things are supposed to be. If we don't, somebody else will.

Our stories become our reality. Since the stories are self-destructive, which is not a healthy thing to be, I think it's fair to call them insane, too. After all, "sane," in its Latin origins, means "healthy," "sound," or "clean."

And it is such insane beliefs that have helped create the world in which we now live.

Dr. Jack Forbes, professor of Native American Studies at the University of California at Davis and author of the brilliant book *Columbus and Other Cannibals,* uses the Native American word "wétiko" (pronounced WET-ee-ko) to describe this collection of beliefs.[2] "Wétiko" literally means "cannibal," and Forbes uses it quite intentionally to describe European standards of culture: we "eat" (consume) other humans by destroying them, destroying their lands, and consuming their life-force by enslaving them either physically or economically. The story of Columbus and the Taino is just one example.

The basis of our culture

We live in a culture that includes the principle that if somebody else has something we need, and they won't give it to us, and we have the means to kill them to get it, it's not unreasonable to go get it, using whatever force we need to. In some cases it's even our *duty* to do so.

"Duty" may seem like a strong word, but it was often invoked by the U.S. government in exhorting pioneers and soldiers to kill Native Americans during the first centuries of this country's history. It was invoked by Hitler to motivate his soldiers during World War II, particularly in the taking of other nations' land for "living space" for the German people. Julius Caesar cited duty as the reason for his soldiers' slaughter of the Celts, Druids, and Picts, among others. Pol Pot invoked duty as his Khmer Rouge soldiers slaughtered over two million of their fellow citizens. During the administration of George Washington, fully 80 percent of the U.S. federal budget was devoted to "Indian warfare." The list goes on and on: for God, country, and family; for Mom and the right to make *your* apple pie out of *their* apples.

In the United States, the first "Indian war" in New England was the "Pequot War of 1636," in which colonists surrounded the largest of the Pequot villages, set it afire as the sun began to rise, and then performed their duty: they shot everybody—men, women, children, and the elderly—who tried to escape. As Puritan colonist William Bradford described the scene: "It was a fearful sight to see them thus frying in the fire and the streams of blood quenching the same, and horrible was the stink and scent thereof; but the victory seemed a sweet sacrifice, and they [the colonists] gave praise therof to God, who had wrought so wonderfully."

The Narragansetts, up to that point "friends" of the colonists, were so shocked by this example of European-style warfare that they refused further alliances with the whites. Capt. John Underhill ridiculed the Narragansetts for their unwillingness to engage in genocide, saying Narragansett wars with other tribes were "more for pastime, than to conquer and subdue enemies."

In that, Underhill was correct: the Narragansett form of war, like that of most indigenous Older Culture peoples, and almost all Native American tribes, does not have extermination of the opponent as a goal. After all, neighbors are necessary to trade with, to maintain a strong gene pool through intermarriage, and to ensure cultural diversity. Most tribes wouldn't even want the lands of others, because they would have concerns about violating or entering the sacred or spirit-filled areas of the other tribes. Even the killing of "enemies" is not most often the goal of tribal "wars": it's most often to fight to some pre-determined measure of "victory" such as seizing a staff, crossing a particular line, or the first wounding or surrender of the opponent.

The European-genocide type of warfare has a relatively brief history, only going back to the days of Gilgamesh. As I've said, it was practiced by Hitler against non-Aryan citizens of Europe, against the citizens of Cambodia by Pol Pot, against the now-extinct Taino and Arawak people of Hispaniola by Columbus, and against the native peoples of the Americas by well-armed invaders from England, France, Portugal, Belgium, Holland, and Spain. It was practiced against the Tutsi by the Hutus in Rwanda, and against the Hutus by the Tutsi in Zaire when it became the Congo. (In the midst of those battles, both groups killed off almost all the three thousand or so remaining Pygmies, the last tribal hunter-gatherers in central east Africa who then lived in the rainforests of Zaire and Rwanda.) There are stories of it in the Bible (see Joshua) and in the histories of nearly all civilizations that have roots in, had contact with, or were conquered by the first city-states of the Middle East.

This type of warfare is practiced daily by farmers and ranchers worldwide against wolves, coyotes, insects, animals and trees of the rainforest, and against indigenous tribes living in the jungles and rainforests.

It is our way of life. It comes out of our foundational cultural notions. So it should not surprise us that with the doubling of the world's population over the past 37 years has come an explosion of violence and brutality. That is our way.

The U.S. government *today* is *still* taking lands from dozens of tribes to transfer their mineral rights to corporate contributors to political action committees; native people are being similarly displaced on every continent of the world, brutally and mercilessly.[3]

But why?

Here are some stories in our culture about "how things got this way."

1. "It's women's fault"

One story, handed down to us by the men who helped build the early foundations of our modern culture, is that it's all the fault of women.

One story says that the first woman, Eve, was stupid and dishonest, a sucker for a snake, and therefore every woman since her has been punished by her god. Or another first woman, Pandora, couldn't control her curiosity. Since women are the cause of all our problems, then it makes good sense—it's our religious duty—to segregate and punish and oppress them. The names of women, with very few exceptions, are discarded from the texts of our modern religions; only the names of men are worthy of being kept as lineage records.

2. "The Creator made us all bad" (a uniquely Younger Culture idea)

Another story is that we're *all* born evil or stupid. Because of Eve's "original sin," every person—man or woman—to pop out of a woman since then is born tinged with evil and bad judgment. This was a story that many cultures, upon contact with Christian missionaries, found particularly difficult to understand. Jonathan Edwards wrote a book in 1822 titled *Memoirs of Rev. David Brainerd: Missionary to the Indians* that Jack Forbes cites in his book *Columbus and Other Cannibals*. In that memoir, Edwards quotes Brainerd as saying:

> It is next to impossible to bring [the Indians] to a rational con-
> viction that they are sinners by nature, and that their hearts are
> corrupt and sinful . . . to show them that they are all corrupt-
> ible and sinful, I observe to them that this may be the case and
> they may not be [aware] of it [because of] the blindness of their
> minds; and there is no evidence that they are not sinful.

As is true of numerous Older Cultures, the Native Americans viewed the concept of original sin and man's sinful nature with bewilderment. Many were equally baffled by the European Gnostic notion of a god who is hateful, vengeful, and delights in setting humans up with temptation and then punishing them when they succumb.

3. "The Creater is a forgetful bookkeeper"

Similarly, many Native Americans thought it odd that our religious leaders said that if a person performed a ritual prescribed by a priest, their god would be induced to "forget" about their having committed murder, theft, or rape. *Wouldn't this allow bad behavior?* they asked. The whole idea of a god whose primary function was to be a book-keeper, but a forgetful one (if the right rituals were performed or words spoken), was incomprehensible to them.

But the idea that everybody is born sinful, that it's our nature to lie, cheat, steal, exploit, and hurt others, is one that some have found a useful rationalization for their behavior and to get others to buy into their worldview. And we've been given further escape hatches: you can gamble the odds on how long you'll live, and then do a penance, give a large gift to charity, or say a particular sequence of words before you die, and you're guaranteed a place in paradise for eternity.

But does this view of the world work?

Consequences of the story that "everybody else is bad, too"

For the spiritual and moral gamblers among us, this story is a license to steal, to do whatever you please.

Consider the American food supply. Plants obtain their nutrients from the soil, and if the soil is lacking in zinc or calcium, so will the plant be, as will the human or other animal that eats the plant. Similarly, as Europeans downwind of Chernobyl found out, if the soil has a sudden attack of radioactive cesium, the vegetables, fruits, and grains will kick over a Geiger counter for years after the harvest. In other words, what's in the dirt is in our food.

So you'd think that the fertilizers they put on our food plants would be pure and clean, right?

Wrong.

Years ago, there were no rules regarding the contents in oil sprayed on dirt roads to keep down the dust, or in the asphalt used to make highways. So, according to an exposé first aired on CBS's *60 Minutes* show, the Mafia in New Jersey came up with a clever trick. Companies that produce toxic waste—lead, mercury, PCB, dioxin, liquid radioactive wastes, the whole gamut—were contacted by a "waste disposal" company that offered to save the toxin producers a fortune by taking their toxic waste for, for example, $200 a pound instead of $3,000 a pound. It was an incredibly good deal for the polluters, and they jumped on it, no questions asked. The "waste disposal" companies, properly licensed and regulated, then took that material and mixed it in with oil, diluting it down to very low levels . . . and sprayed it on rural back roads over large parts of upstate New York and some New England states. They added it into tar to make roads. And in some cases, they even mixed it in with gasoline and sold it to gas stations. The truth came out when a couple of families living along the "dust controlled" back roads became very ill, and the national exposé on *60 Minutes* put an end to the most obvious part of the operation.

That was back in the 1970s. So where did the waste go?

In 1997 a small-town mayor in Washington state launched an investigation into why some livestock in her community were becoming sickly even though they were fed locally grown grains. What she discovered shocked her.

As it happens, the United States has no laws that specify fertilizer cannot contain toxic waste or radioactive materials. So, for example, the uranium processing plant in Gore, Oklahoma, got rid of its low-level radioactive nuclear waste by calling it "fertilizer" (it does contain some nutrients that are good for plants) and spraying it on grazing lands. In Moxee City, Washington, a company accepts toxic waste from steel mills and re-sells it as fertilizer to local farmers. And in Camas, Washington, pulp-mill waste that's laced with lead is sprayed by farmers on their crops. And that's just the tip of the iceberg: this is legal nationwide.

The response of the federal government to this mayor's outrage was twofold. First, in response to the "apple pesticide scare" of a few years ago, pesticide industry lobbyists pushed for and got passed laws in 13 states (as of 2003) that make it a crime to *report a news story* that may cause people to worry about the safety of their foods. (These laws are often referred to as "veggie libel laws," and Oprah Winfrey was prosecuted under the one in Texas for what Texas beef producers interpreted as disparaging remarks about hamburger.) So even the story about the toxic fertilizers never made the pages of some newspapers. Second, the *Seattle Times* found that government regulators were *encouraging* the practice of re-selling toxic waste as fertilizer, because it "saves money for industry" and helps keep expensive-to-maintain toxic-waste landfills from filling up.

Which side is the government on? Well, when *Seattle Times* reporter Duff Wilson asked about why the government would allow toxic wastes to be mixed in with fertilizers and spread on farms and fields across the country, Rufus Chaney of the United States Department of Agriculture told him, "It is irresponsible to create unnecessary limits that cost a hell of a lot of money."

In a similar vein, the U.S. Department of Energy (a euphemistic name for the agency that supervises the production of atomic bombs and warheads) has a great new resource, according to a spokesman for their Hanford, Washington, nuclear waste-storage site. "We have probably the largest stockpile of strontium-90 in the United States," he said in an article published in the July 19, 1997, *Science News*. "Our yttrium [90] is the purest. The process by which it's retrieved from the strontium-90 parent is something that we patented." In that same article, it was revealed that yttrium-90, one of the most common waste products derived from the production of nuclear bombs, is now being tested as a new agent for nuclear medicine, used to attack tumors.

There are a few problems, of course, with using nuclear waste as medicine. An Emory University researcher pointed out that when working with the isotope iridium-192, "its radiation is so penetrating, you would have to move away from the bedside." Nonetheless, the government has high hopes that its bomb-making radioactive wastes can be recycled by being injected into people who may have acquired cancers from eating food laced with other toxic wastes.

Nuclear waste goes to war

A new variation on this took place in the wars between 1991 and 2003 against Iraq and in Bosnia. Over 700,000 tons of "depleted uranium" (D.U.) had accumulated around the world, waste products from the production of electricity in nuclear power plants and the manufacture of atomic bombs.

"Depleted" uranium isn't really depleted of anything—"depleted uranium" is an Orwellian newspeak term used to describe the element U-238, which emits alpha radiation. Its cousin, U-235 emits mostly gamma rays. (Gamma rays easily penetrate flesh, whereas alpha particles are rapidly blocked. This seems like good news, but alpha particles are about 20 times more genotoxic—likely to cause cancer or cell death—than gamma rays. They require closer proximity, however—such as being ingested or inhaled—to assert their radioactive toxicity.)

Both occur together naturally, but the U-235 form, being less stable and with radiation that acts over a large distance, is necessary for atomic bombs and preferred for nuclear power plants. Some of the U-238 in nuclear waste sites was naturally occurring, separated from U-235 in the purifying process to create fissionable bombs and reactor rods. Much of it was the result of U-235 being "burned" in nuclear reactors, leaving a variety of by-products, including U-238.

Lead (atomic weight 82) actually started out its life as uranium (atomic weight 92). Over a period of about 4.5 billion years, half of a pound of uranium will naturally decay into lead as it dissipates energy and mass in the form of rays and particles called atomic radiation. When there's no more radiation left to emit and the nucleus becomes stable at 82 protons and neutrons, one could say that uranium has become truly "depleted" as it has turned itself into lead.

Lead has traditionally been used to make bullets because it's so heavy. Its high mass allows lead bullets to carry more kinetic energy, which they dissipate when they hit an object, causing damage. (This is why you can hurt somebody by throwing a stone at them, but not by throwing a marshmallow. The greater the mass, the more energy and more damage that is conveyed.)

Using lead for bullets has caused generations of soldiers and survivors in conflict areas to be exposed to lead dust and resulting low levels of lead poisoning, leading to small but measurable neurological deficits. Moving from lead to uranium, however, has entirely changed the dangers faced on the battlefield, both by combatants and by later residents of combat areas.

While lead is modestly toxic, it doesn't powder and burn (although it may slightly melt) when it strikes an object, and therefore isn't generally inhaled in large quantities. It's also not radioactive, so although it can damage the nervous system and the kidneys, it's not considered a major carcinogen, mutagen, or factor in birth defects.

Uranium 238, however, burns at about 700° Celsius (a temperature easily created by the energy of a bullet hitting steel), forming a fine dust of uranium oxide, which is easily inhaled and scatters into the wind to contaminate croplands and water supplies. This dust emits primarily alpha radiation—high-speed particles containing two neutrons and two protons—which causes serious damage to nearby tissues and is a well-known cause of lung and stomach cancer.

For example, according to Dr. Helen Caldicott, U-238 in powder form was once suggested as a radiological variation on chemical warfare. "On October 30, 1943," Dr. Caldicott notes, "senior Manhattan Project scientists—the S-1 Executive Committee on the 'Use of Radioactive Materials as a Military Weapon'—in a letter to General Leslie Groves, postulated that the inhalation of uranium would be followed by 'bronchial irritation coming on in a few hours to a few days. . . . Beta emitting products could get into the gastrointestinal tract from polluted water, or food, or air. From the air, they would get on the mucus of the nose, throat bronchi, etc. The stomach, caecum and rectum, where contents remain for longer periods than elsewhere would be most likely affected. It is conceivable that ulcers and perforations of the gut followed by death could be produced."[4]

Facing a huge bill for "disposing" of (storing) millions of tons of U-238, the U.S. and British governments came up with a nifty solution. Because uranium is about 10 percent more massive than lead, it's even more efficient at hitting and killing. It will penetrate through thicker layers of steel than will lead. And it's free!

Although the details are classified as a national secret, it appears that sometime in the 1980s during the Reagan/Bush administration the government came up with the idea of using nuclear waste instead of lead in the rounds (bullets) fired from cannons, helicopters, jet fighters, and tanks. These varied in size from 7.6 mm caliber to over 120 mm. The 30 mm caliber, which can penetrate 7 cm of steel, was heavily used in rounds fired from planes and attack helicopters, particularly against tanks.

During the first Gulf war, over 10,000 rounds of 105 mm and 120 mm were used, and over 900,000 rounds in total. Balkan data, and data from the Iraq War, were not available, although they are assumed to be larger than the amounts first used in Iraq in 1991. Apparently many laser-guided bombs and Cruise missiles also carried D.U. warheads. All told, the U.S. Army disposed of over 300 tons of D.U. by dropping or firing it at Iraq in the first Gulf war, and perhaps three times that much in the 2003 Iraq War.

This has considerably eased the burden of waste disposal for the U.S. nuclear industry, although it may also be responsible for hundreds of thousands of soldiers coming down with Gulf war illness symptoms that are eerily reminiscent of radiation poisoning, and explosions of birth defects and cancers in both Kosovo and Iraq.

How sad it is that we look at situations like this and are shocked but not surprised.

With six billion humans on Earth competing for scarce resources, the notions that "Everybody does it, everybody is sinful, if we don't do it somebody else will, and it's still legal anyway" are pervasive. They're viewed as survival rationalizations. To hell with our children and their world: get what you can *now*! Isn't that the way of our culture, after all? From Caesar plundering the Celts, to Pizarro robbing the Incas, to Columbus enslaving the Taino, to the tobacco industry executives addicting children in Third World countries, it's all the same *wétiko* mind-set: take over another person's life for your own purposes.

Created by a greed-inducing mental illness, culturally contagious, extremely lethal, and rationalized by those who twist religion and culture to justify their own dominating, conquering, and stealing, this mind-set (regardless of the name of the religion or country or culture) is what is killing Earth and her inhabitants.

It's not humankind that is killing Earth. It's the consequence of the *stories* of a now-dominant group of humankind. These stories, which dance through our lives from our earliest childhood and become the lens through which we view other people, other living things, and, indeed, everything in creation including all ideas, are collectively what we call our culture, and what may destroy us if we don't change them soon.

The present story: we're disconnected, separate

The facts that we're running out of oil, that we have less than a 20-year supply of several critical minerals, and that our children may face an economically disastrous future are well known to those who have supervised the rape of the Earth and the transfer of wealth from "natural resources" and native peoples to themselves. The cultural stories that have been used to justify this behavior fall largely into two groups:

- **Get yours before anybody else can.** This story has been with us since the earliest written Younger Culture histories.

   It was the driving force of the theft of the American West from Native Americans between 1820 and 1900.

   During the 1980s it was revived and again became a foundational underpinning for federal government policy. This led to massive consolidations of wealth, a virtual bankrupting of the federal treasury during an orgy of debt-based spending, and the collapse of one sector of the banking industry, as is chronicled in the best-selling book *America: What Went Wrong*.[5]

The "Get Yours" story continues to this day, and forms the classic Younger Culture mind-set as applied to business, government, and personal life.

• **The world is going to end anyway, so grab what you can now.** During his presidency, Ronald Reagan was asked if he thought that the story of the end of the world as played out in the biblical story of the battle of Armageddon was possible. His candid answer was that he not only considered it possible, but that he fully expected it to happen in his lifetime.

This mind-set or story justifies the rape of the Earth by the story that it's all just temporary, anyway. One day soon, fire will fall from the sky, most of the world will be destroyed, and all the Good People will be instantly transferred to heaven.

Notice that these are profoundly disconnected stories: people who live out such stories are disconnected from others (whom they see as either opportunities or competitors), from nature (which they see as a resource to be converted into personal wealth), or from life itself (which they view as merely a game peopled with winners and losers: those who are rich or go to heaven, and those who lost out).

James Watt, the secretary of the interior during part of the Reagan administration, perhaps best exemplified the "I got mine" story. He pushed hard for mining and timber interests to have greater access (and at a lower cost) to minerals and timber on federally owned land because, he said, Jesus will return to the Earth at any time and everything will then be made new, as suggested in Revelation.

Similarly, President Ronald Reagan, Defense Secretary Caspar Weinberger, and right-wing Christian guru Jerry Falwell were quite up-front about their adherence to the "it's all going to end, anyway, so get what you can now" story. Falwell hit the theme of the Rapture recurrently, such as in this sermon: "You'll be riding along in an automobile. You'll be the driver perhaps. You're a Christian. There'll be several people in the automobile with you, maybe someone who is not a Christian. When the trumpet sounds you and the other born-again believers in that automobile will be instantly caught away—you will disappear, leaving behind only your clothes and physical things that cannot inherit eternal life. That unsaved person or persons in the automobile will suddenly be startled to find the car is moving along without a driver, and the car suddenly somewhere crashes."

In February of 1990, Reagan visited Atlanta, where I was then living. In a 30-minute speech (for which he was paid over $160,000) to franchisees of the Days Inn motel chain (who were clients of an advertising agency I started), he said, "It's important to remember that the battle of Armageddon was to begin in an area in Israel, according to the prophecy." Back in 1983, Reagan told a lobbyist for Israel, "You know, I turn back to your ancient prophets in the Old Testament and the signs foretelling Armageddon, and I find myself wondering if we're the generation that's going to see that come about. I don't know if you've noted any of these prophecies lately, but believe me, they certainly describe the times we're going through."

In August of 1984, then-California-governor Reagan told State Representative James Mills, "Everything's falling into place. It can't be long now. . . . Ezekiel says that fire and brimstone will be rained upon the enemies of God's people. That must mean that they'll be destroyed by nuclear weapons. They exist now, and they never did in the past."

The other man with his finger on America's nuclear trigger, Defense Secretary Caspar Weinberger, was even more specific, identifying the Hill of Megiddo in northern Palestine, about 15 miles southeast of Haifa, according to an article in the *Globe & Mail* of Toronto, as the place where the final conflagration that would destroy the world would begin.

These stories that "the end of the world is coming" are not odd-ball, or unusual, particularly among Christians in the American empire. Even among non-Christians, they're fairly common: when former Soviet Premier Khrushchev (an avowed atheist) was asked by American reporters if he thought it was possible that spaceships from other worlds had visited Earth, he said that it was flatly impossible.

The reason, he said, was because any culture that had advanced to the point where they could build spaceships would also develop weapons of war so advanced that they'd destroy themselves.

Whether or not the biblical prophecy of a battle on the plain of Armageddon is going to come to pass is beside the point: I'm not arguing that it will or won't. (If a large enough number of people tell themselves it will, the "field" for it [as British scientist and author Rupert Sheldrake might put it] may become so strong that it would happen anyway.) But with or without Armageddon, our more immediate crisis is caused by people—for over four thousand years—conducting their lives and business as if it's just around the corner. This leads to the get-while-the-getting-is-good mentality that has left our environment and planet shredded.

We need a different set of stories.

Our view of "primitive" people

When we think of the French, for example, we think of people living a relatively happy life, and we have specific color images in our mind's eye—for example, pictures of vineyards and countryside and the Eiffel Tower in Paris. These pictures are probably accompanied by the imagination or memory of sounds—people speaking the French language, having intelligent conversations. And there are most likely feelings (pleasant ones, like curiosity, familiarity, novelty), and tastes and smells (French food and wine, for example).

But when we members of Western Civilization try to visualize and sense primitive peoples like the Kayapo, our mental pictures usually lack color, sound, taste, and smell. Because we've had so little cultural input, or because of the stories we've learned, read, heard, and seen all our lives about the poor lot of primitive peoples, the images themselves are often of people who are sickly, squatting half-naked with stained teeth, scavenging through the jungle, short, poorly nourished, living desperate lives in a hostile environment. If we imagine sounds, it's the sound of meaningless gibberish—certainly these people wouldn't have intelligent, meaningful, insightful conversations.

We may have similar images—or non-images—of the people

who lived before the beginning of our civilization. When we think of "cave men" or "Stone Age people," we imagine cartoon characters, inanimate things that are not quite human and certainly didn't have language, culture, civilization, cuisine, religion, families, communities, or economies. Our images are so gray and distorted because our culture doesn't recognize these "other peoples" as fully human. We refer to that time before the rise of Mesopotamia seven thousand years ago as "pre-history," as if it had no history of its own. It existed only as a footnote to the life of the planet, and if it did have a purpose at all, it was only to set the stage for our arrival.

But the Kayapo and hundreds of other indigenous peoples give the lie to our cultural and religious mythology.

Thousands of years ago, probably even long before the rise of Sumerian "civilization," the Kayapo had a culture that spread across much of Brazil. Skeletons which are thousands of years old indicate that—as with most "primitive" peoples—they had health and a quality of life that was superior to ours, with fewer degenerative diseases; tall, strong bodies; and a long life span. They had a complex spoken language that survives to this day, an ancient oral history, traditions and religions, and covered Brazil with thousands of settlements of up to four thousand people each. These towns were built on huge earthen mounds, which protected them from seasonal floods and created waterways that they used both for irrigation and commerce. They had families, married and cared for children, practiced their religion, and were blissfully unaware of the concept of warfare. (They did have tribal conflicts, but these were never for the purpose of wiping out or destroying another group of peoples. The concept of genocide was alien to them.)

The Kayapo and their peers worldwide, for hundreds of thousands of years, practiced sustainable ways of extracting current-sunlight nutrition, shelter, and clothing from the Earth, in ways that guaranteed their own survival and quality lifestyle as well as that of future generations for as far into the future as one could imagine.

Then Pizarro arrived. Within one hundred years, over 85 percent of the indigenous peoples of South America were dead, most from diseases that the conquistadors brought with them from Europe (influenza, measles, smallpox, plague, etc.). The surviving Kayapo fled into the most interior regions of the Brazilian rainforests and continued to farm with forest-friendly agricultural methods for four hundred years.

Then the loggers and ranchers arrived in the early twentieth century. The woods of the rainforest, particularly mahogany, are highly prized by our culture because of their rarity and beauty. And when the forests are clear-cut, the wounded land is good for pasture for cattle. Ranchers and loggers hired mercenaries to wipe out the Kayapo and other rainforest tribes, offering bounties that were paid when a pair of ears or a scalp was brought back—just as the United States government paid a bounty for Indian ears and scalps in the 1800s. This practice was only recently banned (although many ranchers and loggers still do it, both in Brazil and many other countries around the world). In Africa, for example, the last bounty paid by the government of South Africa for a pair of !Kung Bushmen ears was in 1938. Until that time, hunting and killing Bushmen was considered a "fun sport," drawing hunters from as far as northern Europe and the United States.

That recent ban on killing aboriginal humans in Brazil came about largely because it was discovered that the Kayapo were exploitable in other ways. Their lands have been found to contain valuable woods and minerals, and some companies have made business arrangements with a willing element among the Kayapo people.

The introduction of cash to the Kayapo had many side effects. It "infected" them with our Younger Culture ideas of how humans should live, causing some of their tribes to abandon their traditional agricultural ways and to engage instead in slash-and-burn agriculture to produce cash crops. And so now—as with the natives of other cultures who preceded them, they gave in to "progress," and many are working on corporate plantations and in factories—the exploited have become the exploiters. The Kayapo culture is rapidly disintegrating, as the rainforests are disintegrating along with them.

Our culture's growth has similiarities to cancer

When discussing this recently, a friend mentioned the Gaia theory, which proposes that the planet is a living organism and we're just one cell in that huge body.

Many people—particularly those apologists for the fossil-fuels industry—have taken that a step further and said that since humans are part of nature, then anything we do must be, by definition, "natural." We're nature, after all. This twisted logic is a modern variation of the old rationale of Manifest Destiny.

Consider the medical metaphor as a way of looking at this. Cells within the body are constantly being born, living, and dying. Millions every day mutate in various ways, losing or gaining various amino acids along their DNA chains as a result of exposure to cosmic rays, toxic foods, the by-products of metabolism, and a thousand other natural and unnatural processes. Most of those cells that mutate simply die: their new DNA encoding isn't capable of sustaining life. But occasionally one will alter its genetic code in such a way that the natural switch that controls its reproductive processes is flipped "on" with no way to turn it off. The cell begins to divide—to reproduce— over and over, growing exponentially into a mass of cellular tissue that requires increasingly more and more nutrients. It re-routes blood vessels into it, and consumes the spaces and tissues of nearby organs in its orgy of growth. It takes over. It's called *cancer*.

It's possible to argue that cancer is natural. No doubt there are some cancers that originated from some useful biological process at one time, for example, and there are clearly genes that can make a person more or less disposed to developing cancer. But the majority of cancers are not a normal part of the natural course of human life. They're caused by something assaulting the body, damaging it from without, altering its functions in a wholly unnatural fashion.

To extend the metaphor to man's relationship to the planet, for millions of years the genus *Homo* and for hundreds of thousands of years the species *Homo sapiens sapiens* co-existed with the rest of the planet. Our presence was not unnoticed: we wiped out dozens of species of large land mammals as we traveled from one continent to another, and we altered the natural landscape almost everywhere we went. But we never threatened the delicate web of life, the health of our planet.

Our actions were entirely local. There was nothing that a person could do in 3000 B.C. London when the Druids occupied it, for example, that would affect the lives of the humans living in the Andes mountains on the other side of the planet.

And even our local actions were generally ones that, while they made the environment friendlier to human habitation (herding, growing crops, building shelters), did not exert negative effects on the life of the surrounding areas. The soil was still alive. The forests towered strong and healthy. Animals and plants and fish grew and lived in abundance.

Men and women walked lightly upon the Earth.

A few years ago while I was on a speaking tour of England and Scotland, my hosts took me and my family to see the caves near Chislehurst, outside of London. One large section of the caves—which served as air-raid shelters during World War II—had been carved out of the soft rock thousands of years earlier by the Druids. One area built by them is believed to have been used for religious ceremonies, as there is an altar carved into the wall and the acoustics are so extraordinary that chanting or singing is amplified and echoed to otherworldly effect.

On the Druid altar is a depression, about the size of a modern-day mixing bowl, with a thin groove that extends to the edge of the altar. The British, having spent the period from the Roman conquest to just a few hundred years ago in an orgy of torture, oppression, and domination of women, had recently painted a picture above this spot depicting several women tied into a wooden cage, suspended above the bowl, and being burned to death. "Human Sacrifice!" screamed the poster. "A Druid Religious Ritual!" (Julius Caesar, after his 54 B.C. attempted conquest of the British Isles, wrote that the Druids practiced human sacrifice, burning their prisoners of war in huge wicker baskets. Although this report may be true, it may also be a practice acquired from the Romans, or it may have been one of the ways Caesar justified his ruthless expansion of empire and murder of any race or nationality that stood in his way.)

No one knows for sure what the Druids used the area for. But we do know that they were exterminated by the Romans, the Saxons, and subsequent invaders who now make up most of the population of England. But our hostess, looking at the altar area, pointed out that what little is known of the Druids indicates that they probably were worshippers of the life of the Earth as a feminine and giving power. Their cultural traces indicate that "Mother Earth" was sacred to them, and women seem to have been highly esteemed, as was common among such "primitive" cultures. For example, menstruation, instead of being viewed as the hideous and unclean thing it was characterized as by male-dominated cultures who found women "less useful" when menstruating, was most likely seen by the Druids as part of the monthly cycle of fertility, of life, of nature, and as something to be celebrated or honored.

Looking at the bowl in the altar, our hostess pointed and said, "I'll bet that women would squat over that and expel their menstrual blood, and that would be used both in religious services or to bless the crops. Perhaps it was the menstrual blood of the high priestess, the most holy offering she could make to the tribe, and that's why she and other women were treated with a level of respect that the invading barbarians might have considered 'worship.' "

This seems entirely possible. Riane Eisler and other researchers and anthropologists point to numerous historical female-worshipping or woman-led cultures and societies—some who survived millennia longer than our Western Civilization has, and managed to co-exist with their neighbors peacefully for thousands of years, until they were brought down by Younger Culture invaders.[6]

So to return to the medical analogy of cells in humans, and humans as part of the Earth, historically the genus *Homo* has been part of nature, and has influenced its environment in ways that any other natural predator and modifier of the environment—from tigers to termites—inevitably do.

Assault by Younger Culture leaves one with limited choices

But then about seven thousand years ago, a cancer began to grow. Think of it as an infectious form of civilization, one that continues to be played out even today as children act out the stories and conventions of our Younger Culture.

For instance, I remember in sixth grade a kid named Dennis. He was an absolute terror, and took great delight in physically beating bloody anybody who dared challenge him. Dennis was the undisputed king of the playground, the park, and the paths to and from school and he extracted lunch money from hapless students who encountered him. He embodied the Younger Culture dominator in that little subculture. We knew that he'd learned such behavior from his father, who, according to Dennis, beat him with a belt both for punishment and fun.

Those of us who lived near Dennis and had to pass his house to get to school had three choices: we could avoid him as much as possible, allow him to beat the stuffing out of us, or we could bring weapons or take judo lessons. That last option wasn't available to us in the sixth grade, so most of us chose to elude Dennis as often as possible. Nonetheless, he managed to give me more than one black eye, each time in a humiliating display of his abilities for the largest possible audience.

Those are essentially the same choices available to healthy, sane cultures when they're confronted by a violent dominator culture. The natives of North America first tried to negotiate and make friends with the Younger Culture "visitors" from Europe. When it became evident that these visitors were, instead, thieves and murderers and rapists, stealing the natives' lands, animals, and killing their citizens, some tried to fight back. In order to do so, however, they, themselves, had to adopt those same means to combat the aggressors. And in the process, some became "infected" with the mental disease of domination and aggression, turning into nomadic warriors and hunters of humans.

When two side-by-side civilizations have lived together and traded together for years, and one becomes infected with this Younger Culture worldview, the only choices the other cultures have are to flee, die or become slaves, or adopt those Younger Culture ways themselves.

This is one of the reasons why our Younger Culture is so terribly infectious.

Yet still, there are a few holders of the flame of ancient wisdom, the Older Cultures who may hold in their knowledge, worldview, and history the means for us to survive the next millennium.

Change the story

There is still a possibility we can transform the stories our culture tells itself. This hope is demonstrated in a deep longing that almost everybody in every "modern" society shares.

We see the longing for a reconnection with the world around us in the back-to-nature movement, the idealization of farming, the growing popularity of eco-tourism and camping/hiking, and even in the "sportsman" arena of hunters and fishers. The need for this connectedness is hard-wired into our brains, the result of millions of years of our and our ancestors living in that environment. There is a basic human need for a connection to the environment of our origin, and we can use this need as a lever to change the stories people tell themselves about the world.

Culture is not about what is absolute, real, or true. It's about what a group of people agrees to believe. Culture can be healthy or toxic, nurturing or murderous. Culture is made of stories, and those stories, as we'll examine in the next chapter, can be changed for the better. In changing the stories of our culture, we may find a way to help solve some of the problems we've seen so far in this book.

What We Need to Remember

To plunder, to slaughter, to steal, these things they misname empire; and where they make a wilderness, they call it peace.

—TACITUS (c. 55–c. 120), Roman historian

Gold is most excellent; gold constitutes treasure; and he who has it does all he wants in the world, and can even lift souls up to Paradise.

—CHRISTOPHER COLUMBUS (1451–1506), 1503 letter to the king and queen of Spain

We have lost contact with the memories of our ancient ancestors. We don't remember the stories that were told two hundred generations ago, and, in fact, most of our history books treat that time as if it doesn't exist: it's called pre-history, and in that vacuum of "pre" we have nearly lost the keys to our future survival.

How did we forget? Have other civilizations had similar forgettings? And how can we recover the memories of our distant ancestors?

"Columbus sailed the ocean blue in 1492," goes the school rhyme. Schoolchildren in the United States are taught that this was the time of the "discovery" of the Americas, meaning that it was when the Americas were discovered by Europeans.

North and South America, however, were discovered long before that. Of course, there were the now well-documented expeditions here by Leif Ericson and other Norwegians, and some scholars place Celtic visitors on the shores of New England from around 100 B.C. to just after 54 B.C. This was just around the time when Julius Caesar and the Roman Empire first invaded the British Isles and began killing off, enslaving, and driving out the Celts and their priesthood, the Druids.

But even these possible landfalls two thousand years ago don't mark the first occupation of the Americas by humans: that happened at least 10,000 years ago, and possibly as much as 40,000 years ago.

Current thinking of many paleoanthropologists is that there were three great waves of migration across the area of the Bering Sea, which was a land bridge 10,000 years ago. The earliest of these migrations occurred as much as 35,000 to 40,000 years ago, and led to the settlement of the Arctic by the people who now call themselves the Inuit, and were referred to in the past by Europeans as Eskimos. The second migration may have brought people as far south as what we now call Argentina, leading to the populating of South America possibly between 25,000 and 15,000 years ago. (There is an ongoing debate about this: some believe that these people crossed the Pacific by boat about that time.) The third migration occurred around 10,000 years ago, and led to the human colonization of North America.

In 1492 Columbus thought of himself as the representative of one great empire (that of the European nations in general, and Spain in particular), on his way to visit another great empire (India). In actual fact, however, the two largest empires in the world at that time that were known to the Europeans were the Ming Dynasty of China and the Ottoman Empire of Turkey. They controlled more than half the known world between them, and had so locked up the trade and resources of Asia and much of Europe that one story of Columbus's journey is that he was off in search of trade routes to India that wouldn't require passing by Turk-controlled waters or Ming-controlled land routes. If he could find such a route, the Spanish could avoid paying tolls and taxes to these great empires, thus vastly improving the profitability of their trade. (The idea that people in his day thought the world was flat is a simple myth: every sailor knew the world was round, as they could see ships vanish over the horizon. It was common knowledge, both in academic and sailing circles, and had been for hundreds of years.)

Another speculation is that Columbus was actually in search of America—more specifically in search of gold and slaves that could be taken from America—as there were numerous reports of others who had discovered "that distant and rich land to the west," including Portuguese expeditions in 1460; numerous Basque trips between 1375 and 1491; and a departure from Bristol, England, in 1481, returning in 1491, which reached the Newfoundland coast. The word was in the air, and no queen had to hock her jewels to outfit Columbus.

However it came about, what Columbus discovered on Hispaniola was infinitely more profitable than tax-free trade with India: he found slaves and gold. Columbus returned home a fabulously rich man. This led to a gold- and slave-fever explosion in Spain and Portugal. Within 30 years, by 1520, Spain had landed numerous forces along the beaches of Mexico, and extracted tens of thousands of pounds of gold from the natives.

About that time, one of the Spanish conquistadors heard rumors of a great empire far to the south, a land where the buildings were faced with gold and the people were fabulously wealthy. In 1532, Francisco Pizarro led a force of 260 mercenaries to the coast of what is now Peru. With 62 horsemen and 198 foot soldiers, he disembarked and traveled upland into the Andes to the Inca town of Cajamarca, where he requested an audience with the Incan emperor. The emperor traveled to Cajamarca for the supposedly peaceful meeting with the foreign visitors, and Pizarro captured him and his entourage, holding them ransom until several weeks later when he was paid two roomfuls of silver and one roomful of gold (over $60 million at today's cost of gold and silver bullion).

This ransom payment convinced Pizarro that there was something to the tales he'd heard about the Incas, and so he garroted the emperor and his aides, buried their bodies, and began the long march to Cuzco, the mountaintop capital of the Incan Empire of Tahuantinsuyu. What he found was the capital of the world's largest empire at that time, a nation unknown to the rest of Europe and Asia, but more populous than Ming China or the Turkish Ottoman Empire, and far larger than Spain.

The Incan Empire ruled over what is now Peru, Argentina, Chile, Colombia, Bolivia, and Ecuador—almost all of South America except for the jungles and rainforests of what is now Brazil (which the Portuguese later claimed). The Inca had built more than 40,000 kilometers of all-weather highways, a road system that was unequaled anywhere in the world until the advent of the automobile. Their empire was divided into 80 political provinces, and, like the Romans, they had imposed a lingua franca on all the disparate peoples they ruled, Runa Simi being the required tongue.

The city of Cuzco was, indeed, studded with gold. There were huge esplanades, sparkling fountains, massive buildings for administration and governance, and majestic temples. And gold gleamed everywhere—gold ornaments worn by the citizens, entire gold walls on the inside and outside of the temples and royal palaces. Gold statues and figures of the various Incan gods, particularly Inti, their god of the sun, filled the city and its buildings.

Like the Roman Empire and the European empires that followed it, an elite family ruled the Incan Empire. Numbering fewer than 40,000 people, this clan comprised the only "Incas" in the empire—everybody else was a serf or slave or villager. The Incan royal family emerged about the same time the European royal families emerged—between 600 A.D. and 1000 A.D.—and, like the European royal families, had extended and consolidated their rule by 1500.

That Pizarro was able to read his famous decree, conquer the largest empire in the world with only 260 men, and ship back to Spain hundreds of tons of gold was viewed by the Spanish as a sign of divine providence. The reason the conquest was so easy was that when Pizarro arrived in Cuzco in 1532, already more than 60 percent, and perhaps as many as 90 percent, of the population of the Incan lands was dead.

The unintentional introduction of smallpox to Mexico by a Spanish conquistador in 1520 led to a plague among the Native population—which had no immunity whatsoever to the European disease—that ran through the native populations of Central and South America like wildfire. By 1524, smallpox had killed almost the entire population of Panama, and once it jumped the isthmus it spread across South America, killing almost everybody in its path.[1]

Wayna Capac, the last Incan emperor, died of smallpox in 1525, along with his son and heir apparent, and most of the rest of his family. The power vacuum and demographic collapse was so great that when Pizarro arrived seven years later there was only a sickly remnant of the once-mighty civilization left to attempt to oppose him, and a non-family member had assumed the role of emperor.

The Incas left behind so much gold, particularly under the ground in their elaborate and multigenerational burial chambers, that the Spanish government titled out the Incan lands of South America as mining stakes. By 1537, the gold rush was in full fury with the arrival of tens of thousands of Spanish, and the Castilian (Spanish) king had established an official smelting operation in the Moche Valley. This wasn't to smelt gold from ore (which the Incas had been doing for almost a thousand years), but to melt down the hundreds of thousands of gold items being brought in from the looting of the mausoleums of Chan Chan and the stripping of the Pyramid of the Sun. They melted these priceless artifacts down so the gold bullion could be easily shipped to Spain, and the king's men supervised the operation so they could collect his 20 percent tax. To this day, hunting for Incan artifacts is a major (albeit technically illegal) underground industry among the poor living in most of the former Inca lands of South America.

"The Great Forgetting"

Australian Geoff Page and artist Bevan Hayward (whose Aboriginal name is Pooaraar) produced a beautiful book of photographs, poems, and stories titled *The Great Forgetting*.[2] The book's title is a phrase the Aborigines of Australia have long used to describe what has happened to their culture as a result of two hundred years of forced assimilation into the European-based white Australian culture. More recently, author Daniel Quinn used the phrase in his books *Ishmael* and *The Story of B* to describe the phenomenon of a conquering/assimilating culture both destroying and forgetting the remnants of the origins of their own culture as well as those that they had assimilated.[3]

The average citizen of South America today, regardless of ancestry, has little interest in, and even less knowledge about, the culture and lives of the people who lived on that continent before the arrival of Pizarro. The time of the Incas is forgotten, part of "pre-history," not even taught in Peruvian history classes, but instead consigned to the arcane realms of archaeology and paleoanthropology.

The people were vanquished, most of them dying from smallpox and later under the sword of the invaders, and their gold and other valuables were taken. And then they were forgotten.

But that wasn't the first Great Forgetting.

When survivors of the Inca slaughter by the Spanish were interviewed around 1530, they said that the Incas were the first civilization to arise on the South American continent. The sun god Inti placed the first Inca man and woman there, they said, and from them arose the nation. The lineage was known, who was the son of whom, all the way back to the original man and woman.

But while that's what the average Inca believed in 1530, it wasn't the true history of the area. In the north highlands of Peru, for example, the Inca ruled from approximately 800 A.D. to 1500 A.D.

From 400 A.D. to 800 A.D. the Marcahuamachua peoples controlled the area.

Before them, back to around 10 A.D., the Recuay Empire ruled the region. They were preceded by, respectively, the Chavin, Kotosh, Huacaloma, and Galgada empires, which first arose about 2000 B.C. Before the Galgada there were the Lauricocha people from 8000 B.C. to 2000 B.C., and the Guitarrero people lived in the area from 10,000 B.C. to 8000 B.C.

The Incas forgot all of them, just as most modern Peruvians have forgotten the Incas themselves.

Even though the Incas first appeared 15,000 years ago, they practiced the techniques and had the culture of domination, oppression, and genocide. They built their empire by conquering nearby peoples and turning their citizens into slaves. Their empire was made up of a small ruling elite, about 1 percent of the population, who controlled well over half of all the nation's wealth, and perhaps as much as 90 percent of it. In these aspects, they were not all that different from the Spanish who conquered them, or from contemporary Western/European/American cultures now. All are Younger Cultures.

On the other hand, some of the Older Culture people who had survived the onslaught of the Incas still survive today. For example, the Kogi of Colombia continue to view the soil, oceans, rivers, forests, and sky as living and sacred. They viewed with horror the conquest and domination of peoples and lands by the Incas 15,000 years ago, as much as they view with horror our rape of the planet today by the descendants of European culture. They know that others preceded them, that the Earth's history is extraordinarily ancient, and that the planet will continue on with or without us.

The beauty of remembering

My mother is fascinated with genealogy. She's traced our ancestors back to President James Madison, and in the tenth century, to the original Prince of Wales (before the British royal family usurped the title when they conquered the Welsh). I feel connected to Norway, the country my father's parents came from just before World War I. When I read the histories and family trees she and other members of our family have unearthed and organized, I get a feeling of belonging, a sense of history, a sensation of continuity and groundedness. I wonder how those people lived their lives, what they did, how their thoughts and actions may be still echoing through my DNA and through the world. I study what I can of their goals and ideals, and those help form or reinforce or cause me to question my own values.

This sense of history is an essential one for humans. It's critical to a healthy culture, which is why we make the study of history mandatory for schoolchildren. It's important for self-esteem, which is why so many American blacks are pushing for a non-European view of Africa and the history of slavery. It's why almost every religious or political leader has tried to either re-write or carve their place in history (or both).

And yet our view of history is peculiarly short-term and narrow.

In the cultures formed out of Jewish, Christian, and Islamic religious foundations, it is taught that the original Hebrew tribe began with Adam and Eve, and their surviving sons, Cain and Seth. They were put on the Earth about five thousand years ago.

The Genesis story refers to other humans ("in the land of Nod") with whose daughters the sons of Adam and Eve mated and produced offspring, but these people were members of other tribes, and so they're mentioned only briefly.

The way the story is told serves to cut us off from memory of the other humans alive before the birth of the gatherers, Adam and Eve, their farmer son, Cain, and his herder brother, Abel. It erects a memory-wall to anything preceding that time.

Dominator Younger Cultures work best when their members believe that they are unique in human history, they are continuous with the First Man and Woman, and that they are chosen by the universe's Creator to rule over all other cultures (an assumption implicit in the first two assumptions).

Such cultures will fight to the death to preserve these assumptions, and will take "whatever necessary steps" to either kill off or create amnesia in the citizens of other cultures that may challenge this belief.

When Pol Pot murdered between one and two million Cambodians, he did so to begin "a new society" from a clean slate. I met Pol Pot's personal physician, Red Cross doctor Will Krynen, in Indonesia about 12 years ago, and he told me how Pol Pot believed that if he could induce a complete amnesia among the populace about their history—by killing off anybody who could read or write, and those old enough to remember the past—he could build a new society by raising the surviving children with a "new past," which he invented, designating himself as the father of their culture and civilization.[4] For that reason, he ordered that the year following his major slaughter of Cambodians be designated the year "0." All calendars in Cambodia would henceforth originate at that date, and the history books written for the young Cambodian survivors of the slaughter referred to the time before the year 0 as "pre-history," having little relevance or importance, and mentioned only in the most vague of detail. (Consider why we number our calendar as we do now.)

Pol Pot had studied history carefully: he knew that others before him had done exactly what he envisioned, and succeeded. He nearly succeeded, and probably would have if the Vietnamese hadn't intervened by invading Cambodia to stop the slaughter.

The founders of modern Japanese culture understood this lesson well, too. In the eighth century, their foundation thinkers developed an elaborate story that the first emperor, Jimmu, derived from the sun goddess, Amaterasu, in the year 660 B.C. and created the unique Japanese race. This story was taught as fact in Japanese schools up until 1946, and so fervently believed that Kamikaze pilots enthusiastically gave their lives for the god-descended emperor.

I was astounded when I first visited a Native American tribe and discovered that until just three decades ago it was the official policy of the U.S. government to imprison Native Americans who practiced their own religion. They were not allowed to perform certain dances, have sweat lodge ceremonies, sing certain songs, or pray in certain ways, and those who were caught performing these "heathen" religious rituals were put into prison alongside murderers and thieves. It is *still* the policy of the government to imprison "unapproved" Native Americans whose religious practices include eating shamanistic, mind-altering plants that naturally grow on their lands and which have been intrinsic to their religious practices for thousands of years before Europeans set foot on this continent.[5]

Similarly, anthropologists studying Native religions and culture from the times before the arrival of Columbus, particularly in Central America where there is much interest in the Mayans, are having a particularly difficult time. The Catholic Church decreed the Natives "heathen" and therefore the Spanish went on elaborate "search and destroy" missions to find any and all art, records, temples, hieroglyphs, or anything else that may allow the Mayans and others to remember their past and continue their culture. Their language was forbidden, their religion was condemned, and anybody found practicing either was put to death. (It was the same process Caesar followed in the conquest of Europe, when he destroyed the many tribes living there.)

In the United States, the Spanish hadn't had a chance to be as thorough as they were in South and Central America, and many Native tribes were still intact when our soldiers and settlers began moving west in the 1800s. When the annihilation of the Native Americans didn't work (or was stopped), we enforced laws for centuries that were designed to create amnesia among them, to strip them of their identity. Many of these efforts were led by the Catholic Church, which still runs schools and other programs on many Indian reservations.[6] The practice is not unique to America. In Australia, for example, it's just in the past two decades that the government has stopped the practice of forcibly taking part-Aboriginal children from their parents and placing them in white foster homes so as to cause them to forget their Aboriginal culture.

Because of the way dominant culture educates the masses, the average citizen of the modern world knows little to nothing about this. The prevailing notion is that primitive people are . . . well . . . *primitive*. Even that word, which for several hundred years our ancestors applied to the aboriginal peoples of North America, was openly used to imply inferiority, hunger, a rude set of semi-functional social skills, childishly simplistic technology, and a laughably naive religion. The most well-known Native American in twentieth-century American culture is the fictional character Tonto, sidekick of heroic cowboy the Lone Ranger. "Tonto" is the Chiricahua Apache word for "slow" and Spanish for "dumb."

It's only since World War II, when a segregated group of Native Americans were allowed to fight on behalf of the Allies and ended up highly decorated, that Americans have begun to experience respect, guilt, intimidation, and even awe when they meet Native Americans who are still living their Older Culture lifestyle.

In fact, as any careful read of the literature of anthropology or a visit to tribal people will tell you, the depth of human experience is no different between "primitive" and "modern" people. Both have identical ranges of expression and emotion, both have cultures that are clearly defined with standards and norms for behavior, both have rituals and religions that are deeply meaningful to their citizens. The prime differences are that the "primitive" people generally have more leisurely lives, less poverty, almost no crime (certainly no police or prisons among those who have not adopted "the white man's ways"), a more diverse and healthy diet, less degenerative disease, better psychological health, and a culture that holds as its primary values cooperation (rather than competition), mutual respect (rather than domination), long-term renewable care for resources (rather than exploitation for a quick buck), and equality (between people, between the sexes, and between humans and nature) rather than power.

Anthropologist Mark Nathan Cohen, in his book *Health and the Rise of Civilization,* points out that over the past 30,000 years of the well-researched human fossil record, it's only in the past hundred years or so that agricultural peoples have had life spans that exceeded those of hunter-gatherers and foragers.[7]

The record, in fact, is startlingly clear: 30,000 years ago the average height of an adult human male was 5 feet, 11 inches, and the average height of a female was 5 feet, 6 inches. Men in agricultural communities, starting 10,000 years ago and extending up to just the past 200 years, averaged only 5 feet, 6 inches, and women had shrunk down to a mere 5 feet tall.

Thirty thousand years ago, the average adult died missing only 2.2 teeth; by eight thousand years ago in agricultural societies it had risen to 3.5 teeth missing, and by Roman times tooth decay had exploded to where the average person died missing 6.6 teeth.

And this was not because people were living longer: in fact, the average male life span of 33.3 years found during the Upper Paleolithic period was only again approached among agricultural societies when, in the United States in 1900, the average life span of non-white males hit 32.5 years. (Since that time, it is primarily antibiotics that have accounted for the sudden increase in First World life spans since sulfa drugs were first developed during World War I and penicillin during World War II.) In general, foragers and hunter-gatherers ate a healthier, more varied diet, got more human-appropriate exercise than did agriculturists, and lived less stressful lives in greater harmony with their environment and their neighbors.

As *Columbus and Other Cannibals* author Jack Forbes points out, it's more than a little ironic that people we call primitive and uncivilized had evolved a way of life that worked so well that they didn't *need* police or prisons. Since reading that observation of his, I've noticed that there's a sure way to tell how unequally a society divides its assets: the more concentrated the wealth and the more violent the society's dominators, the more prisons there are.

What we must remember: the "Older Culture" view

Forbes points out that, with few exceptions, most Native American cultures did not have our notions as part of their collective mythos. Instead of the story that we're "separate from creation and born to dominate it," these cultures hold a different view of the place of humans in the order of creation:

- **We are part of the world.** We are made of the same flesh as other animals. We eat the same plants. We share the same air, water, soil, and food with every other life-form on the planet. We are born into life by the same means as other mammals, and when we die we, like them, become part of the soil that will nourish future generations.

- **It is our destiny to cooperate with the rest of creation.** Every life-form has its special purpose in the grand ecosystem, and all are to be respected. Each animal and plant has its own unique intelligence and spirit. We are permitted to *compete* with other plants and animals, but we may not wantonly *destroy* them. All life is absolutely as sacred as is human life. (Even using the word "sacred" here is difficult, because it implies that something else is "not sacred." In these Older Cultures such distinctions do not exist. Life is, and that is a thing of extraordinary importance, at the very core of all existence.) Although hunting and killing for food are part of nature's order, when we do so it must be done with respect and thankfulness.

Older Cultures are most often *cooperators*, not *dominators*. *There are human cultures who do not engage in the destruction of the world. They demonstrate that destruction and domination are not an inevitable part of human nature.*

Prior to the emergence of Younger Cultures about seven thousand years ago, the anthropological record shows that not one culture believed itself to be separate from and superior to nature. We find the remnants of these Older Cultures in tribal peoples around the world such as the San, the Kogi, the Ik of Uganda, the Navajo, the Hopi, the Cree, and the Ojibwa—living in harmony with the world around them, the people around them, and seeing all life as sacred. The San bushmen don't even qualify as Stone Age since they've never used stone implements, only tools made from wood, and yet they were successfully pursuing their way of life 40,000 years (and perhaps 100,000 years) before Aristotle. They leave behind few traces, as they are such masters of resource management.

That's sustainable and, contrary to the stories of our culture, it was and is often a happy and comfortable life.

When we lived like that thousands of years ago, we enjoyed cradle-to-grave security. The tribe took care of itself, we cared for one another. If anybody had food, everybody had food; if anybody had a diseased child or an infirm parent, everybody had a diseased child or an infirm parent. The measure of wealth in such societies was security. Mediums of exchange like money were unnecessary; the idea of hoarding food or other things was unthinkable, because everybody was responsible for everybody. Our ancient ancestors lived in the way of all other cooperator societies in nature, be they the society of wolves or chimpanzees or prairie dogs: they looked out for one another.

Our ancestors—people like you and me, of all races on all continents—lived like this, all over the world, for 40,000 to 200,000 years, depending on whose archaeology you accept.

And then there were eruptions among traditional cultures. In some parts of the world, people began to move away from their hunting and gathering lifestyle by experimenting with agriculture. This created more efficient food production, thus increasing their numbers, and giving some people the ability to hoard food: the beginning of "wealth." (Today we use money to try to *buy* that cradle-to-grave security that *all* of our tribal ancestors enjoyed as a right of birth, but only a very, very few of us ever achieve it.)

Then a sub-group of the agriculturists began experimenting with a new cultural idea of coercive or forced evangelism, of bringing others into their culture in a way that had never been done before. Their gods told them that if they couldn't evangelize others, then they should utterly destroy them. They were a very few (probably not more than a dozen) tribes, which arose out of the tens of thousands that populated the planet, and this small number of tribes proceeded to wipe out and displace and destroy the thousands of other tribes who were living in a sustainable, peaceful, and connected-to-nature way. They left the Garden and began to create dominating city-states and then empires.

The birth of class differences and power structures

They were the first people infected with *wétiko,* the origin of our Younger Culture, and because of this they became more efficient at increasing their own numbers.[8] They had more sunlight under their own personal control. (I say "infected" because Younger Culture is contagious: people who are attacked by Younger Cultures have few choices, and those who survive have most often become Younger Culture themselves.)

Of course, there was a price to pay for this. While the San, Kogi, Ik, and other Native peoples may spend less than two to four hours a day gathering food and attending to the needs of life (and do to this day), in Younger Culture societies this balance was radically shifted as "average people" must work longer and harder just to survive. Those who were the dominating individuals in the culture, however, could live luxuriously and work less and less.

So for every person who only worked an hour or two a day, another person would have to work four or eight or ten hours a day or more. Without massive exploitation of resources or theft from others, for every person with ten times as much wealth, ten people must have only a tenth as much. Social and economic classes were born, and the first governments came into being to define, order, and control the socio-economic structure, and help the wealthy maintain and increase their riches.

Whether they knew it or not, these governments—mostly kingdoms in the early days—transmitted Younger Culture values to all citizens, rich and poor. The power brokers of this time "programmed" the consciousness of their subjects, just as our governments, educational institutions, and mass media do today.

How it happened

Nobody knows what brought about the first eruption of *wétiko* cultural insanity, but logic suggests that it most likely happened in places where food resources were cyclically abundant. For example, the Tlingit and Waida Native American tribes of the Pacific Northwest in the area around Vancouver Island were apparently extensive traders and owners of slaves long before the arrival of Europeans (who were also slave owners). According to anthropologists (who may be contaminated by Western bias), as many as 25 percent of the local population may have been slaves at any one time, with 7 percent to 15 percent being the norm. Why?

Some anthropologists theorize that because the salmon run twice a year in this area and produce a wildly abundant but brief food source, these tribes developed ways to store salmon (drying, salting, etc.) to equalize their food supply year-round. This equalization of food also had the novel effect of providing a much larger year-round food supply, which meant that the local land could support much larger human populations. And, in fact, the average tribal unit in many of these areas ran into the hundreds of individuals, whereas the hunter/gatherer tribes farther inland had tribal units that rarely exceeded 50 to 100 people. Apparently this same small tribal pattern prevailed in Europe from the first arrival of humans between 60,000 and 40,000 years ago until the emergence of "civilization" 5,000 to 10,000 years ago, and we still see remnants of it among the Laplanders of northern Sweden, who are the last indigenous people of Europe.

Along with the opportunity to support more people, however, the ability to preserve food created a second culturally destructive side effect. This storing of food created the first wealth.

Those who excelled at storing food, or at stealing stored food, ended up with the greatest food-wealth. During food shortages, individuals or tribes had to submit themselves to the will of the wealthy ones in order to obtain enough food to survive.

This storing of food was perhaps the first human step away from nature.[9] It created a separation between humans and the natural world. Accompanying this was what turned out to be a self-destructive arrogance and belief that nature could be dominated, eventually leading to the idea that other people could be subjugated or exterminated.

The "slavery" (losing your freedom) of civilization

In 1861, Mark Twain rode the railroad and overland stage across much of the United States, documenting his journey in *Roughing It,* published in 1871. During a stagecoach ride near the Great Salt Lake, he encountered a group of the Shoshone-speaking Native Americans called the Gosiute, often then referred to by whites as the "Digger Indians." Twain considered them "the wretchedest type of mankind I have ever seen," and wrote, "[they] produce nothing at all, and have no villages, and no gatherings together into strictly defined tribal communities—a people whose only shelter is a rag cast on a bush to keep off a portion of the snow, and yet who inhabit one of the most rocky, wintry, repulsive wastes that our country or any other man can exhibit. The Bushmen and our Goshoots [*sic*] are manifestly descended from the self-same gorilla, or kangaroo, or Norway rat, whichever animal-Adam the Darwinians trace them to."

Even to this day, many people who have not bothered to study the Shoshone or other hunter-gatherer people imagine them in much the same way Twain did. Books and movies over the years have implied that their lives must be a continuous wretched struggle to find food from day to day, and that their culture and religions hardly qualify as either.

In that belief, however, Twain and many modern people are wrong. If the highest goal of contemporary civilization is to have leisure time, free from the demands of providing for food and shelter so that one may then contemplate the great mysteries of life, then the Shoshone had achieved the pinnacle of success!

Our culture teaches that civilizations (city-states) come about as a result of technological innovations (such as agriculture) giving people more free time. With this free time, the story goes, they produce art, literature, religion, and explore the cosmos. "Primitive" cultures don't have these things because they don't have the time for them.

In fact, these represent two of our most deadly myths.

Leisure time

Every empirical study of both historic and contemporary cultures finds that the more complex and hierarchical a culture is, the harder the people in it must work and the more frantic their lives are. Just look at how many hours a week the average middle-management executive works (about 60), and how many families have two 40-hour-per-week workers devoting 80 hours a week to paying the mortgage and feeding the family.

Only a very small class of people within the city-state enjoys the "leisure time" state of "freedom": its economic and political rulers. And, because the ruling class is not producing food, those who are food producers must spend extra time making food for those who are not.

The Shoshone require the same average 2,000 calories of food energy per day as do any other humans. However, they expended on average only two hours per day to acquire it, because they were a nomadic people who moved from place to place following their food supply. As the seasons changed and food became scarce in one place, they simply moved to another. If one food wasn't available, they knew where and how to find another.

Toronto University's Professor Richard Lee found that a similarly structured tribal group, the !Kung of the Kalahari desert in Africa, spent less than 15 hours a week (about 2 hours a day) attending to gathering food and other necessities of life. The rest of the time, he said, they played, told stories, and made music. John Yellen of the

National Science Foundation found the same to be true of the Hottentots, another hunter-gatherer group in Africa.

Depth of culture

The Shoshone had an elaborate and meaningful culture and religion. They generally did not suffer from famines or plagues. They had lived comfortably and happily on their land for at least several thousand years, and perhaps as many as 10,000 years, keeping the land as clean and pure and productive as it could be in that desert and mountain region, living harmoniously with their neighbors.

At the time Mark Twain took his ride through their territory, the Shoshone had accomplished—over a thousand years—a second achievement, which is regularly touted by our leaders as the highest goal of humankind. They had eliminated warfare. There was not even a word for "war" in their language.

The Shoshone lived a tribal life in one of the most desolate parts of North America, with a population density that ranged from one person every fifty square miles to one person every hundred square miles. A typical tribal unit was a single extended family of five to twenty people, and they traveled at a leisurely pace across a wide area. On those rare occasions when others (including whites) came to attack them, they simply ran away and hid. The occasions for attack were rare, largely because the Shoshone accumulated no wealth: they had no systems for preserving and storing food, minerals, or anything other than what they could carry. In this regard they were not poor: their lives were comfortable, their family interactions meaningful, and their food supply ample. A symbol of this is found in the highest-status act a Shoshone could commit in the presence of others: to give them what he had. Generosity was how one achieved social standing among the Shoshone, whereas the accumulation and control of surplus food and possessions were how whites achieved social status.

They were called the Diggers by whites because they often dug in the ground for roots and food. Whites assumed this implied some sort of agricultural stupidity, but in fact the Shoshone had a deep and rich knowledge of life in their environment both above and below the ground. They used a sacred digging stick to extract food, and it was both manufactured and transported with ritual and ceremony. If a stone had to be moved, a different type of stick was used. When a Shoshone looked out at the natural world, she saw a landscape rich with life, both visible and hidden. That life was known to her, called to her, spoke with her, and often guided her.

The Shoshone culture was filled with rituals and rules that were, to quote their chronicler, the late Peter Farb, "every bit as complex as those of the Vatican or the Court of Versailles."[10] Throughout their lives, they had to be aware of the spirits of nature and the worlds beyond nature that surrounded them, to monitor their interactions with others for appropriateness, keep a mental record of obligations and past interactions with family and other clans, and know where the sacred and profane places were so they could be visited or avoided. A particularly elaborate etiquette surrounded their rites of passage, including marriage, birth, death, and puberty.

Shoshone life was largely egalitarian. Leadership was an advisory capacity, and depended upon ability. The best hunter led the hunt; the wisest and most experienced medicine man or woman was the group's physician; the best gatherer of foods led expeditions to search for plants. Since these levels of knowledge and experience changed from person to person as they went through life, the persons exercising a leadership role changed often as well. They viewed leadership as an obligation, rather than the opportunity for power and wealth that it's often viewed as by "civilized" persons. It carried a heavy burden, and so was treated with respect and often shared among several people. It was not aspired to or worked for, but instead almost "inflicted" on the most competent by the rest of the tribe. The fluidity of leadership among the Shoshone was a characteristic that was particularly confusing to European whites when they first met them.

While an Iowa corn farmer today must produce 12 million calories of food a day with his two thousand calories of "life energy" expended in being alive (and is only able to do so because of oil-driven technology), a Shoshone had only to produce four thousand calories of food a day. This is because of what anthropologists refer to as cultural overhead. The more energy a society puts into creating non-edible "things"—be they cathedrals or toys or living spaces—the more energy those who do produce food must produce with their efforts. While our cultural overhead is massive, that of the Shoshone was relatively modest: the extra calories produced by the adults were largely used to feed the very young and very old among them.

This is also why the Shoshone rarely experienced famines: they didn't have a huge structure of production and storage that could be tipped over. When food supplies became thin in one area, they simply moved to another.

In all these regards, the Shoshone, like most other small-scale tribal people, were remarkably free from the burdens of any form of slavery. No person "worked" for another, none was "owned" by another, none spent his or her time producing food for anybody beyond his or her own immediate family group. They devoted an average of two to four hours a day to finding food, and had the rest for leisure and ceremony. (This ratio is typical of tribal peoples the world over.)

Modern-day slaves

In modern society, few people report that they feel even remotely "free" in our modern society: we are modern-day slaves, held captive by "slave-holders" of our culture. The slave-holders use the chains of the mortgage owed the bank, the loan on the car, the unpaid credit card bills, the requirement to pay property taxes if you own your own home, and the many other subtle and not-so-subtle forms of economic and cultural pressure to extract the majority of your life's time and use it to their ends.

As a result, almost everybody in modern society knows somebody who's on tranquilizers or has lost control to alcohol. Addiction to television is so rampant it is causing the disintegration of traditional social groups and clubs, and our children are lost in a sea of pain and confusion that has led to a *doubling* of the rate of teenage suicide in the past three decades.

Slaves know when they are slaves, regardless of the words used to describe their slavery. And they'll seek escape from the slavery, be it in increasingly powerful drugs, increasingly intense "entertainment," or psychopathic or violent behavior.

We must begin to teach our children and our citizens how to search for a more true history of the world, and encourage them to look for the truth of the present. Only then can we re-connect with our past, and thus begin to create greater personal identity, collective identity, and collective responsibility.

From this new sense of who we are and what our place is in the world, the things we have to do to help save the world become both apparent and possible: without this perspective they appear overwhelming and impossible.

In my explorations of this field, I've concluded that the Older Culture peoples of the world have important lessons to teach us. Indeed, they may well be the lessons that will save our world. . . .

The Lives of Ancient People

 The mission of the United States is one of benevolent assimilation.

—U.S. PRESIDENT WILLIAM MCKINLEY (1843–1901)

From the San and the Kogi: value community and cooperation; we are part of the world, not separate from it

One of the oldest cultures on Earth is that of the !Kung Bushmen of the Kalahari desert in the northern parts of South Africa. The exclamation point in their name !Kung represents a sound in their language which we don't have in English: it's a popping noise made in the mouth by forming a vacuum between the tongue and the top of the mouth and then pulling the tongue down quickly. There are three other sounds in their language for which we have no letters, all of them clicks or pops, made by similarly clicking the tongue against the front of the mouth or the sides of the mouth and teeth. They are such a unique culture that although they're ancient, their language contains sounds that have traveled to no other human tongue on Earth.

Over the past few decades, as they've become more well known, they've asked anthropologists and linguists that they be called the San, although most texts from before the 1980s refer to them as the !Kung. (They and their life are portrayed wonderfully in the film *The Gods Must Be Crazy*.)

The San are racially distinct from the other Africans who have conquered the continent in the past few millennia. Their skin is more yellow than black, and their eyes are slightly slanted, as if they share a common ancestor with Asians, or perhaps are indeed an early ancestor of the Asians. Their hair is black and curly like other Africans, but they're comparatively short and thin, often standing less than five feet tall and weighing less than one hundred pounds.

Laurens van der Post, a South African explorer and writer, first chronicled the lives of the San, quite elegantly. In his 1961 book, *The Heart of the Hunter,* he tells of coming across a small !Kung tribe of about a dozen adults and children as they crossed a particularly hot and barren part of the desert.[1] Van der Post and his fellow explorers started hunting some game so the Bushmen could have extra food to carry on their journey "toward the lightning on the horizon" where the seasonal rains were beginning. The explorers spent an entire day hunting with their Land Rovers and provisioned the Bushmen well for their trip.

As the little tribe was leaving, van der Post and his group stood to wave good-bye, but the Bushmen simply walked off with many smiles. No thank-yous were ever given. One of van der Post's assistants, a hunter who'd never encountered Bushmen before, commented that they seemed ungrateful and uncaring. Ben, one of the other men in the group who understood Bushman culture, responded that to give another human food and water is only good manners and is routine behavior among the Bushmen. If the white men had been starving on a long trek and the Bushmen had found them, they would immediately share their food and water, even if it endangered their own survival. They wouldn't expect thanks in response.

In fact, in San Bushman culture, to eat in front of another person who is without food is an immoral act, every bit as horrific as in our culture if a person were to walk out onto a busy city sidewalk, pull down their pants, and defecate. Everybody would be shocked and horrified.

As it happens, the San do say "thank you." They do it whenever they're hunting, when they're making a decision to take a life. No animal is killed for food by the San without being thanked by them, both at the time of the hunt and later when a dance is done for the soul of the animal. Also, animals are killed only when there is a clear need for the food.

For those of us who grew up in modern civilization, it's difficult to imagine a life and culture where such fundamental things are taken for granted. When we stop behind a car at a red light, we don't open the door and run up to the car in front of us to thank them for being so considerate as to follow the basic rules of the road and stop for the red light—it's a given that everybody does that. No thanks are required. Thanking people for doing something implies that they had a choice to do otherwise, and did it out of a desire to be nice.

But imagine a world where feeding another person is as much an automatic response as stopping for a red light; a world where a person who fails to feed or care for another is ostracized or punished, the way we give people tickets if they run red lights; where the care of others is more important than even the care of yourself; where the teaching "All things that you would want others to do to you, do ye even so to them" is actually practiced—not out of effort but as part of the daily routine, as the normal way things are, as a basic assumption of society.

That is San culture, the way of an Older Culture.

A storyteller of Chippewa and Cree ancestry told me that his people have a belief that if a person visits your home, and you fail to share with them food and water so that they leave hungry or thirsty, and then the Creator decides to "take them home at that time," they will arrive in the Spirit World hungry or thirsty. "The responsibility for that, for that person's condition in that world, is yours, because you were the last person he met and you then had an opportunity to feed him. So we have an obligation to feed and give water and shelter and whatever else a person may need whenever they come into our village or our home."

In our Younger Culture, we value productivity and individual possession. In their Older Culture, they value community. Most "modern" people find it difficult or impossible to imagine a world where community is more important than possessions, yet this is how about 1 percent of the world's population still lives, and how all of your and my ancestors lived for 100,000 years.

In 1997 a group of 13 researchers released a study in which they quantified the value of all the environments of the planet. From measuring the size of Louisiana shrimp harvests to how much people were willing to pay for access to a lake, coral reef, or other natural attraction, they concluded that the planet's natural areas were worth about $33 trillion.

That someone would even consider putting a price tag on the world is an indication of how far out on the edge we've gone. It demonstrates a mind-set that says that the world is here for us and has value only to the extent to which we can or do use it. According to this perspective, "natural resources" are only a "resource" if they are usable by humans.

Many people share this viewpoint. From those who claim that the planet is a self-stabilizing, living system to those who argue that we need more wild areas to preserve forests for campers and back-packers, the implicit message is that we need to save ecosystems *because they are of value to humans, directly or aesthetically.*

There are those who wax poetic about the views from the Pacific Coast, or the astounding vitality of Amazonian rainforests. We have to save these environments, they say, so that our children and their children can appreciate them. Or we need to save them because those trees are the lungs of the planet and that shoreline is where unique life-forms exist that may one day be discovered to have the cure for cancer. Keep it because we may someday want or need it.

The Kogi Indians, however, look out on the mountains of South America's Sierra Madre chain, the Great Mother of All Life, and see that while Mother has provided a place for humans, their "younger brothers" of our Younger Culture are now on the edge of destroying Mother herself. Our jets pierce her like so many needles, criss-crossing the sky; we dig into her flesh and tear out her innards with our mining equipment; we drill deep into her and drain out her fluids with our water and oil wells; we throw soot and waste and smoke into her face and onto her body. The Kogi have sent out emis-saries to tell the modern world that they are horrified at what they are seeing: we are killing the Mother of All Life.

Even at its most noble, its most altruistic, its most concerned for our environment, our Younger Culture is expressing a profound self-centeredness, a concern that if our natural environment is lost, we may no longer use, appreciate, or even worship it.

In all cases, what's implicit in our cultural view of the world is a hierarchy, a good-better-best and a bad-worse-worst. Nature is better and more noble than humanity, or humanity is superior to nature and has a noble obligation to subdue and exercise dominion over nature. Good guys and bad guys.

There's a different way to view the natural world. Older Cultures, with few exceptions, hold as their most foundational concept the belief that we are not different from, separate from, in charge of, superior to, or inferior to the natural world. We are part of it. Whatever we do to nature, we do to ourselves. Whatever we do to ourselves, we do to the world. For most, there is no concept of a separate "nature": it's all us and we're all it.

From the Kayape: sustainable agriculture

The Kayapo are a Gê-language-speaking tribe of native peoples who live in the rainforests of northern Brazil. They've been there at least two thousand years, and many researchers believe they've lived in that area for as long as eight thousand to ten thousand years. Their way of life has been continuous for that entire time . . . until recently.

The Kayapo practice an interesting form of agriculture, based on the idea that you can take what you need from the forest or fields, and even manipulate the forest and fields so they produce more human foods and medicines, but that you cannot do this in a way that injures the land.

They begin by creating what are called circular fields. Starting at one particular point in the forest, they'll fell trees in a 10- to 20-foot area, with each tree falling so that its crown points out toward the edge of the circular clearing. This produces an open area covered with felled trees, which radiate out from the center like the spokes on a wagon wheel.

In the first year, they plant legumes and tubers such as manioc, potatoes, and yams, among and between the felled trees. These plants stabilize the soil, and many of them fix nitrogen and other nutrients into the soil. At the end of the growing season, the Kayapo burn the trees, distributing the ash around the soil to fertilize it. The burning doesn't hurt the root vegetables, which are then dug up and stored or eaten.

In the second year, edible plants are sown in circles from the center of the clearing out toward the surrounding forest. The plants that have the greatest need for sunlight, such as sweet potatoes and yams, are in the center, then progressively more shade-loving crops are planted—corn, rice, manioc, papaya, cotton, beans, and bananas in rings that move toward the outer periphery. The most shade-loving plants are on the outer circles.

Each year a new field is prepared and, around the seventh year, the first field is abandoned for agriculture so that the forest can reseed it and new trees begin to grow in the still-fertile soil. Many of the crops continue to grow wild in the area—particularly the potatoes and yams—and are harvested for years as the forest reclaims the field. For the first 10 or 20 years as the field returns to forest, berries, medicinal herbs, and small fruit trees proliferate, providing a new and different food source. There is also a lot of bush and underbrush that grows up, providing home for small game, which the Kayapo hunt to supplement their diet. Within 20 years, the area is once again rainforest.

This sustainable agriculture enabled the Kayapo to build a huge culture over millions of acres of Brazil before the invasion of South America by the conquistadors.

How different their world—and that of the San—is from ours.

Power vs. Cooperation in Social Structure: the City-state vs. Tribes

 Every gun that is made, every warship launched, every rocket signifies, in the final sense, a theft from those who hunger and are not fed, from those who are cold and are not clothed. The world in arms is not spending money alone. It is spending the sweat of its laborers, the genius of its scientists, the hopes of its children.

— President and five-star general DWIGHT D. EISENHOWER (1890–1969), April 16, 1953

Recently I heard a self-appointed prophet preach about how the world was going to end soon because the god of his sect was angry with humans, particularly those of another political party. "Two-thirds of all people living will die!" he shouted. "Plague and famine and fire from the sky will kill them!"

My first thought was that the sudden death of two-thirds of humanity would mean utter cataclysm. It would be difficult to find enough burial space for all the bodies, while the stench and disease risk would be unimaginable. Stacks of corpses would line city streets, as they did in 1350 London during the bubonic plague epidemic that killed half of that city. The New York Consolidated Metropolitan Statistical Area (CMSA, a Census Bureau term for the aggregation of contiguous metropolitan areas) would have more than 13 million dead bodies in it, reducing its population from 20 million to 7 million; the Los Angeles CMSA would decline from 15 million to 5 million; Chicago from 8 million to about 3 million. The 95 million people of Mexico would become just over 30 million; Italy's 57 million would drop to 19 million; China's 1.2 billion, to 400 million. The world would seethe with dead and dying people, and those left would live a nightmare existence.

My second image looked entirely different. If two out of three people died today, those left would still be more numerous than the entire population of the world in 1930. If that preacher's numbers were in fact conservative and instead five out of every six people died, the Earth would still hold as many people as were alive in 1800. Imagine, if 23 out of every 24 people alive today were to die, there would still be more people alive than at the time of Christ, and the world was by no means sparsely populated then.

Cycles of boom-and-bust, rise and fall, over-use of resources and then overwhelming scarcity—even cycles of famine and plague—are *normal* for population-dense, growth-and-consumption-based city-states. They have always happened in the past, first on local levels, then regionally, then nationally, as we've seen with the great empires of the past. If enough nations confront them together, they could occur globally.

Such cycles rarely, however, happen to people who live tribally, maintaining their lives on the sustainable base of local resources. The reason is intrinsic to how they're organized, a fact well known by America's founders such as Thomas Jefferson and Benjamin Franklin, who wrote extensively about the importance of maintaining local economies and democracies.

Tribal and city-state cultural structures

There are two basic social organizations of humans that we know of: *city-states* and *tribes*.

Tribes have been around for the entire 100,000 years of known human history: the smallest tribal unit is the family, the largest historically has been fifty to a few hundred people. (Some groups that are really city-states call themselves tribes, such as the contemporary Zulu of Africa, but they are organized as a city-state and not a tribe.)

Tribes have historically been highly successful human systems.

From the dawn of human history until seven thousand years ago, tribes were the human race's sole representatives over the entire planet. As recently as 1800, tribes populated half the Earth.

The structure of a tribal or democratic group

The evidence from analysis of tribal peoples alive today is that tribal life is relatively stress-free, satisfying, produces more leisure time than city-state life, and—perhaps most important—is sustainable indefinitely.

Tribes are characterized by five primary traits:

1. Political independence
2. Egalitarian structure
3. Getting their resources from renewable local sources
4. Having a unique sense of their own identity
5. Respecting the identity of other tribes

Political independence

A tribe is a politically independent unit, usually numbering between a few dozen and two hundred members. Of the most nomadic people (like the San), a tribe may be far-flung, although the various families that make up a tribe usually live in close proximity.

One of the problems the early settlers had in dealing with the tribally living Native Americans is that the settlers, based on their own stories about social organization, expected to find a hierarchical city-state organization (local groups like towns, larger groups like states, and so on) where none existed. For example, the settlers would negotiate a deal of some sort with a local tribe of 30 to 50 people and then assume that the deal applied to all Native Americans who called themselves by the same name or spoke the same language. But that was and is not the case: there were thousands of Cheyenne, Apache, and Paiute tribes, each a politically independent unit.[1]

Egalitarian structure

Leadership in a tribe is an advisory role, not an authoritarian one. (There are exceptions to this, but the anthropological record shows that they are rare.)

Again, early European invaders of the Americas didn't understand this; in fact, they considered it a sign of backwardness, and so sought out the "chief" or leader of a tribe, thinking that they could negotiate with that person and everybody else in the tribe would have to comply. In fact, tribal leadership is usually held by a committee, and even that committee is more advisory than authoritarian. Power is shared among the members of the tribe, as are resources.

As the kibbutz movement in Israel has shown, this variation on communalism works in small "tribal" groups; as the experience of the city-states of the former Communist world has demonstrated, communalism fails in the larger city-state systems. Modern people, viewing the world from their own perspective, assume there must be high-status persons in tribes, such as medicine men, shamans, chiefs, etc. A careful read of the record of people who have interacted with tribes prior to their widespread contamination by dominator city-states shows that people with these titles are considered equal with everybody else, and view their role as carrying an obligation of service, not an opportunity for domination.

Locally based, dependent on renewable local resources

Tribes eat what is in their area. If that changes or goes away, they move on. Some tribes have regular traveling areas, spending a few months to a few years in one area before moving on to the next, allowing the first area to regenerate. Others are stable or agrarian.

The two key concepts here are "local" and "renewable." Tribes live in intimate contact with their local environment, and so develop religious and social/law systems that emphasize the importance and value of the natural world. (Because tribes develop and husband resources for the long term, their living areas are—unfortunately—generally rich in natural resources and therefore attractive to the predators of city-state cultures.)

A sense of unique identity

A member of a tribe is born into that tribe. The tribe defines his or her identity. Tribes do not evangelize (go out trying to get others to convert to their ways), do not accept "converts" or "new residents," and are convinced that their way of life, their stories of the world, and their gods are the best *for them*. An Apache, for example, would no more think of declaring himself a Cree than he would of declaring himself a wolf or a mountain. This ethnocentrism works well, in that it guarantees the long-term survival of the tribal unit. Similarly, the diversity among human societies guaranteed by many different tribes make for a stronger overall human presence: diverse ecosystems are strong, whereas single-species (or, in this case, single-culture) systems are fragile and prone to collapse in upon themselves when stressed.

Respect for other tribes

While tribes occasionally compete or come into conflict, they most often cooperate, as seen in the rituals of the potlatch and pow-wow. One tribe may view another with disdain for their social, religious, or other practices, but there are few historical records of tribal people engaging in genocide. Another tribe, although different, may be fought with, but never wiped out entirely. After all, the other tribes are useful. They produce different goods, which can be traded for. They are genetically diverse, so intermarrying with them (usually on ceremonial occasions, or as part of trade economy) guarantees a strong gene pool. And, perhaps most important, by being a "them" the other tribes help a local tribe to maintain its identity as an "us."

Although intertribal conflicts sometimes result in deaths, it's rarely a large number, and in the case of most tribes studied over the years, intertribal conflict usually produces no deaths. The function of the intertribal conflict is to help solidify and maintain the boundaries and unique character of each tribe. As such, it's a good thing for the survival of both tribes.

The structure of a non-democratic city-state culture

About seven thousand years ago, the first politically organized city-states came into being. Since that time, they have systematically exterminated almost all remnants of the tribal cultures they come in contact with. This process of extermination is now nearly complete; this century has seen the extermination of more tribal people than any in history. Brazil alone exterminated 87 tribes between 1900 and 1950, and today tribes represent only between 1 and 2 percent of the total human population of the planet.

The story our culture tells itself about the destruction of tribes is that primitives must pay the inevitable price of progress. Darwin and Thomas Huxley implicitly proposed that tribal die-off was a natural process, demonstrating the inherent superiority of the city-state form of social organization over the tribal. They were "primitive" and we are "advanced," and so for the process of natural selection to work as it must, sooner or later they will be gone. The same thing has happened to thousands of other plant and animal species in the past: if they couldn't survive, they were wiped out, and the world is the better for it.

Now, however, we're beginning to see the flaws in a city-state organization.

1. Because pre-democratic city-states are hierarchically organized, there is a concentration of power. In a Younger Culture, this results in a concentration of wealth and the existence of have-nots.

2. Being immersed in such an organization leads us to assume that all of nature is organized hierarchically, and that our place is at the top. This assumption makes it seem reasonable to engage in actions that foul and destroy the "inferior to humans" rest-of-the-world.

3. The result of these assumptions is that we've wildly exceeded a sustainable population, damaged our atmosphere, and endangered our food and water supply, and we are producing microbes that are more deadly to our species than anything ever imagined by our ancestors.

4. Pre-democracy city-states have always had a history of rising up to dominate for a (relatively) brief period of time, and then collapsing.

Contrast tribal characteristics with the structure and nature of city-states:

1. Political dominance
2. Established hierarchy: clear authority structures
3. Acquiring resources through trade and conquest
4. Absorbing other cultures into their own identity
5. Genocidal warfare against others

Political domination

While the city-state/nation/kingdom itself may claim political independence, on the level of individual citizens and families there is no independence. The local units of family and local community are dominated by the larger political entity of the city-state. This creates a mind-set of domination and hierarchy that we see played out in businesses, families, local communities, and the organized religions that nearly always arise from city-states in forms that serve the ends of the city-state.

This is most obvious to Westerners in the old European model, where the king owned all the land, crops, animals, trees, and even the people—and used political domination, armies, police, torture chambers, and prisons to force his subjects to give him a share of their lives or production. (Well known in Europe but less well known in America, this extended even into the most intimate moments of the citizens' lives: "The Rite of the First Night" was standard fare in Europe for 1,600 years, where every woman who married had to spend her first night after the wedding losing her virginity to the king or local lord, and then joined her husband the next day. This practice is first documented in the *Epic of Gilgamesh*.)

While modern forms of political oppression are more or less overt, depending on the country, the principle is the same: the citizen exists to serve the government, and must give that government a portion of his or her life, time, or wealth on a regular basis.

Hierarchical, not egalitarian

Pre-democratic city-states are organized in such a way that the most powerful, aggressive, or wealthy individuals rise to high positions, whereas those with little power, wealth, or willingness to be aggressive sink to low-status and low-power positions. This type of internal social organization is one of the engines that drives city-states to expand constantly, as the most powerful and wealthy individuals aggregate and consume more and more of the resources available to the city-state. This leaves less and less for those on the bottom, driving the demand for growth to avoid unrest or revolt.

Trade- and conquest-based

Self-sufficient, locally based city-states are rare, although Jefferson and others of the Founders argued strongly that all communities in America must ultimately be self-sufficient if they are to remain democratic. Smaller nations that are self-sufficient are often easy targets for those that are not and so the larger ones trade or conquer in order to maintain their continual expansion: such movements have politically remapped the world a dozen times over in just the past few generations. Because of their hierarchical social, political, and economic structure, city-states must acquire many of their resources from outside sources in order to maintain their growth. When they've exhausted local resources (as seen in the history of nearly every city-state), they attack, conquer, or absorb their neighbors. This process continues as neighbors become more and more distant, until the entire planet is consumed . . . at which time the city-states may begin to collapse—just like a Ponzi scheme, or a company that's run out of its startup capital. A culture that depends on conquest for survival is not sustainable when the limits of worldwide resources are approached.

Exploitative evangelism and "absorptive" identity

Growth is the prime directive of city-states. When growth stalls, they often collapse politically, socially, and economically, or are conquered, or internal power is seized in coups. Because of the importance of growth, they've adopted several methods to expand.

The first is absorption of other peoples and their resources. Slaves were brought from Africa to Europe and the Americas; as one European nation-state conquered another, they brought their new "subjects" under their rule. The Native Americans were conquered and their resources appropriated by the European/American city-state/nation. The results of this were an expansion in population, increased productive capacity and consumption, and more consumers for the output of the city-state.

A second way city-states grow is by assimilation: converting people from other tribal identities to theirs. Evangelists convince tribal people that their way of life is bad or sinful, and give them opportunities to "join" (albeit at the bottom of the hierarchy) the culture/religion of the city-state.

While tribal people would never evangelize, this is a cardinal characteristic of most city-states, and has been enforced by the threat (and action) of death, torture, wholesale extermination (as we saw during the Crusades, the Inquisition, and the Conquering of the American West, and now see in the enslavement of indigenous peoples of South America and Asia), or afterlife damnation.

Warfare with other city-states

Because growth is a high value for city-states, they come into conflict with other city-states (or tribes) who have resources the city-state needs. While city-states may maintain a dynamic equilibrium for some time, appearing to be stable (like the United States and Canada), history shows that this stability is relatively short-lived. Eventually, a city-state's consumption will exceed its ability to produce local resources, and it will have to begin to consume the resources of others. It may use the weapon of food/money/resources to do this, as illustrated by the fact that over 70 percent of the fruit in U.S. supermarkets comes from Third World countries, or it may use weapons to assert its (or its proxies' or allies') claim to land, people, and resources, as we saw during the Gulf War.

Once others willing to engage in warfare confront a city-state, its choices are very few, just as we saw with a culture being confronted by *wétiko*. Survival typically involves becoming just like the attackers: becoming good at warfare.

How city-states might have started

Somewhere back in pre-history, a tribal leader violated the tribal worldview, or became insane (by the definition of his people). He defied the tradition of cooperating with nearby tribes, and instead plotted to conquer one and turn its people into slaves. Perhaps, in getting his tribal members to cooperate, he used an incentive similar to that used by Columbus. He allowed his men free reign to rape any of the other tribe's women they wanted, and even take young women as slaves to clean and perform sexual services. Or perhaps he used the technique of Pizarro—who ordered his ships burned when he arrived in the Americas so that the faint of heart couldn't desert—which ensured complete and absolute domination of his tribal members.

Perhaps this first *wétiko* tribal leader had a rationalization or what he considered a justification, such as a weather change producing poor crops in his area, or a lack of game, so that his people were hungry. Or perhaps he convinced his people that the gods had talked to him and given him this terrible command. However it came about, he and his tribe attacked and conquered a nearby tribe.

Warfare and genocide were invented.

In the process of conquering the nearby tribe and either exterminating or enslaving their people, he discovered that by using force against others he instilled fear in his own people because of his willingness to use violence. Once his own people were afraid of him, they did whatever he asked, whether it was to join his band of killers, pay him part of their hunt or crops, or give him their children for labor or warfare.

Dominating, fear-based leadership was invented.

Taking by force a portion of each tribe member's productive output or personal goods added to his power. He was able to share some of his surplus goods with those closest to him, and they, in turn, helped him solidify and maintain his domination.

Wealth and the use of capital were invented.

He looked at women, who in most tribes hold places of high esteem because of their ability to bring new life into the world, and realized they represented a threat to his new style of leadership. He pointed to their menstrual blood, which had been a sacred substance to sprinkle on fields or use in fertility ceremonies for tens of thousands of years, and called it unclean.

He pointed to the pain some women experienced during childbirth, and said that his god had told him that it was a punishment. The women were wicked, the consorts of lesser or evil gods. They had the power to make men want them, and because he knew so well how corrupting power was, he decreed that women must be controlled, hidden, dominated, their social status reduced from equal to that of property. When crops failed, or people died of illness, or natural disasters occurred, it was the fault of the women and their witching ways.

Sexual domination and patriarchal hierarchy were invented.

His people looked to the night sky and the forces of nature—lightning and earthquakes and wildfires—and concluded there was an omnipotent power afoot, which caused things in individuals' lives to go well or poorly for its own purposes. He told his people that the gods had chosen him as their spokesman. He invoked sacred names and powers, and his ability to dominate in warfare was seen as proof of the blessing of the most powerful of the gods. He forbade people to worship any god but the one who talked through him, and sent out emissaries to either convert or kill those people who didn't bow before the god that spoke through him. Those who agreed to believe his words, he allowed to join the tribe, once they'd sworn their loyalty to him and his god.

Exploitative and enforced evangelism was invented.

Viewing other humans as objects to be dominated, it was a short step to view the natural world as something to be dominated. Instead of following the renewable agricultural practices that had kept his tribe alive for tens of thousands of years, he decided to extract as much food as possible from his land, regardless of the consequences. When the land became exhausted, he needed to take somebody else's land, and could, now that genocide and slavery were available tools. If any other species competed with his people for their food—wolves that ate their sheep, or small animals that ate their plants, or even insects—he would do what he could to destroy those "enemy" species.

Scorched-earth agriculture was invented.

* * *

Tribal history worldwide is littered with stories of bands who became insane in one of these ways because of hunger or lust for power and killed off their neighbors. Some tribes grew beyond normal tribal size and had access to seasonal food supplies, thus creating hierarchies of wealth and power. Other tribes engaged in non-sustainable scorched-earth agriculture, and either killed themselves off or were forced to move to new lands. Still other tribes believed that their gods were the only true gods and all others were either less powerful or altogether false. None of these renegade tribes, however, ever rose to conquer the known world. This is because never before had all these elements come together in one place. If they had, dominator civilizations would have emerged at that time.

Until, that is, seven thousand years ago, when one man became the first dominator, the first evangelist, the first scorched-earth agriculturist, and, because of this aggregation of what tribes consider at least three individual types of insanity, became the first builder of a city-state.

History indicates that it may have been King Gilgamesh in the Fertile Crescent area of the Middle East. Most likely, though, Gilgamesh was only a descendant of the first man to invent our culture; Gilgamesh merely fine-tuned this new synthesis of social elements in such a way that he could arise to conquer and deforest his entire known world, and write his own history of it.

The tribal peoples that Gilgamesh and his cultural descendants confronted had no defense. As his new form of social organization touched the Syrians, the Greeks, the Romans, the Hebrews, the Arabs, the Vikings, the Turks, the Huns, the Imperial Europeans of Britain, Germany, Spain, France, Portugal, Belgium, and Holland, the "Americans" and "Australians" who came from Europe, the Incas, the Bantu-speakers, the Zulu, the Chinese, the Japanese, the Koreans, the Brahmins of India, and others conquered and transformed every tribal people who stood in their way.

Many tribes lacked the central power structure that would make warfare something they could fight. They lacked the degree of job specialization that could produce weapons of war and standing armies. They lacked the willingness to destroy every competitive living thing within their environment in order to extract the maximum possible amount of food from it. They lacked the belief that their god wanted them to kill others and would bless them with happiness if they did so.

They were utterly unprepared. And so they fled into progressively more and more isolated and less productive lands until at last they had nowhere else to run and hide. And those who weren't killed off were "assimilated."

The story of the Toradjas tribe is a good example, and fairly typical. The Dutch had "conquered" the Celebes Islands (now known as Sulawesi), and there lived in the Poso district of these islands a hilltop-dwelling people known as the Toradjas. They grew a dry variety of rice, and hunted, gathered, and lived tribally. Their economy had no money or other means of exchange beyond social courtesy and obligation, and hunger was unknown to them. They were quite happy with their lifestyle, which they had maintained even thousands of years before Holland first was occupied by dominators from Rome, and they had no particular interest in planting crops for export to Holland or in working for the Dutch lowland owners on their coffee plantations.

This situation was intolerable to the Dutch, who observed that under such circumstances "development and progress were impossible" and unless something was done quickly these tribal people were "bound to remain at the same level" of primitive lifestyle. So in 1892, the Dutch governor sent in missionaries to destroy tribal culture. This effort, however, was a total failure. Even offering "free education" in the mission schools for the Toradjas' children wasn't enough to convince them that they should give up their religion or way of life. They simply had no interest in buying goods from the Dutch-owned stores, or in planting and growing coffee or rice for the Dutch export business, or in worshipping the gods of the Dutch. Without cheap native labor, the local Dutch industries were hardly as profitable as they could be.

After 13 years of diligent effort by the church, the Dutch government implemented Plan B. They brought in the army, and forcibly relocated the Toradjas from their ancestral lands on the hilltops to the lowlands. They took Toradjas men for slave labor (they called it conscription) and used them to build roads, then imposed a head tax on each of their citizens. In order to pay the tax, the Toradjas had to go to work in the coffee plantations, and by 1910 they were "converted," sending their children to the mission schools, buying Western clothing and appliances, smoking tobacco and drinking alcohol, and adopting Christianity. Although their mortality rates had soared, and they'd exchanged the healthy, leisurely life that was lived by their ancestors for 10,000 years for one of frantic and grinding poverty, they were now, according to the Dutch government, "civilized."

This same scenario has been played out literally thousands of times across Asia, Africa, Australia, and, of course, North and South America. In some cases, well-meaning donors even send money to support programs to "save the heathen" of distant lands; this is happening with increasing velocity in the jungles of Brazil and Southeast Asia, for example, where "civilized" interests want the natural resources of the jungle and need the natives for labor.

The third and final option available to tribal people is to fight. If they cannot run and hide, and prefer not to be "assimilated," then they must engage in battle. This is particularly destructive, because it requires that they first must adopt the culture of their enemy. To organize an effective army requires hierarchical social structure, specialization of labor, and dominator leaders. Resources must be consumed at a frenetic pace, leading to a loss of the quality of life and often causing hunger and poverty. At that point, before the first shot is fired, the Older Culture has already lost the war: they have become their enemy.

Tribal populations

We see an interesting pattern when we look at tribal civilization versus city-states. While, like cancer, the population growth of city-states has never historically been controlled or controllable other than by plague or famine, the populations of tribal peoples tend to remain stable over thousands of years. We are taught that this is because they had such poor sanitation or unreliable food supplies that they experienced high infant mortality and a short life span.

Recent discoveries show this is not the case.

In the era prior to antibiotics, tribal people generally lived longer than city-state dwellers (on average), and had lower rates of infant mortality. Further, studies of fossils of tribal people compared with early city-state dwellers shows that the tribal people generally had fewer dental caries, stronger bones, and fewer signs of degenerative illness. Many paleoanthropologists over the years have called the agricultural revolution and the creation of city-states a public health disaster.

Steven Mithen, in *The Prehistory of the Mind,* states that the paleological and contemporary records clearly show that "the onset of agriculture brought with it a surge of infections, a decline in the overall quality of nutrition and a reduction in the average length of life."[2] Why, then, did humans develop agricultural communities? Mithen points out that while agriculture led to deterioration in the quality of life for human societies, it also "provides particular individuals with opportunities to secure social control and power." Pointing to Darwin's theories of natural selection, which propose that evolution works more often to the benefit of the individual than to the group, "we can indeed see agriculture as just another strategy whereby some individuals gain and maintain power."

But how do tribes control their population?

One of the things that causes modern people, residents of city-states, to believe the story of the exploiters that it's beneficial to "save" the tribal peoples of the world by destroying their culture, is the question of population. How do tribes control their populations, if not by cannibalism, infanticide, rampant disease, or high levels of infant mortality? While our culture assumes these are their methods of population control, this is not the case. In fact, populations in modern developing nations such as Mexico have higher rates of infectious disease, infant mortality, suicide, murder, malnutrition, and hunger than any known indigenous tribal group ever studied. Yet tribal populations tend to remain relatively stable for, literally, thousands of years. How do they do it?

Nobody knows.

In *Victims of Progress,* author John Bodley points out that the significance and mechanism of how tribes control their populations "is still not fully understood," but that it is definitely not anything from the list above.[3] Numerous excavations of ancient tribal people have yet to demonstrate any evidence of widespread infanticide or even higher levels of infant mortality than are found today in the majority of countries in the "modern" world. They simply don't overpopulate, and nobody knows exactly why.

One theory is that fertility is a function of available food supplies. While we haven't studied human populations in this way, we do know that both wild and caged animal populations will always grow to the capacity of their food supply to sustain their population, *and then stop.* (Fish in tanks will even grow their bodies to an appropriate size for the tank and then stop. Nobody knows how.) It may be there are subtle internal biological and endocrine feedback systems that tell the body if there is or is not ample food or living space. When the food supply drops below an optimum level, the endocrine system is triggered to reduce sperm count or motility, or the viability of ova, or even to release fewer hormones and pheromones that stimulate a desire for sex.

Another theory has to do with exercise. In a study published in 1997, it was found that 57 percent of women who were cross-country runners had amenorrhea, a condition in which menstruation stops its normal cycle and a woman is temporarily infertile.[4] While amenorrhea is viewed as a "disease" of exercise by modern medicine, which attempts to treat it by returning menstrual regularity with the addition of estrogen and other hormones, in a natural environment it may have been part of a delicate mechanism to balance tribal populations. If more than half of all women on the planet for the past five hundred years had been infertile at any given time (taking turns, as it were, as their levels of exercise changed in response to the need to help gather and hunt food during lean times), then we might not have the population explosion we are now experiencing. Similarly, a 1993 study found that the likelihood of a woman getting both breast and ovarian cancers was *increased* by early-adolescent-onset menstruation (exercise among adolescents moves back the date of menarche), late-life-menopause, and, perhaps most significantly, regular menstruation.[5] The more frequently a woman menstruates, it appears, the more often she is exposed to body hormones that can promote the growth of these types of cancer. Among highly active women, there are fewer overall menstrual cycles during the lifetime, and thus less exposure to estrogen, the hormone known to promote some types of cancers.

A third theory is that although tribal people seem ignorant by our standards, when it comes to the things that directly affect their lives, they are often advanced far beyond many modern peoples. Tribal people had been using penicillin, for example, for thousands of years before its "discovery" by modern scientists around the time of World War II. They'd been using the Pacific yew tree to treat breast cancer for five thousand years before the "discovery" of Taxol in that plant in the 1990s. Similarly, many plants are now known to contain compounds that directly affect estrogen or other hormonal functions in both women and men. The chaste tree (*Vitex agnus-castus L.*), for example, was used for thousands of years in Europe, then adopted by Greek "pagan" medicine women to reduce men's sex drive so that the goddess of fertility wouldn't be offended during the feast of Thesmophoria and the crops would grow. Its names derive from *hagnos* and *castus,* both Greek words for chastity. Other herbs, such as tansy and rue, are such effective abortifacients, or "morning-after" drugs, that they were commonly described as such in medical textbooks up until the early years of the twentieth century. So it may well be that people with an advanced knowledge of natural pharmacology (as is seen in every tribal group studied) would use their knowledge to control their fertility.

Two other variables that reduce the rate of population growth and are easily observed among tribal people are breast-feeding and homosexuality. It's common for tribal women to breast-feed their children for as much as three to five years. During this time, the body produces hormones that inhibit menstruation and fertility, presumably to prevent the woman's body from being stressed by the double-whammy of breast-feeding and growing a fetus at the same time.

While cultures that encourage large families—often to produce more formidable armies—usually develop religious and cultural injunctions against homosexuality, such taboos are often conspicuously absent among tribal people who don't have the imperative of producing more and more cannon fodder. As well documented in *Living the Spirit: A Gay American Indian Anthology* by Will Roscoe and *Spirit and the Flesh: Sexual Diversity in American Indian Culture* by Walter Williams, gay and lesbian people and activity were generally both accepted and even celebrated in many Older Cultures.[6] Having as much as 10 percent to 20 percent of the population engaged in non-procreative sexual activity would also have a stabilizing influence on population, and when population pressures increase, more people may become or be born homosexual (this is what happens with most mammals—the first rat experiments to demonstrate this were conducted 50 years ago, and have been replicated many times).

A final theory is that in most tribal cultures, women hold positions of status and power equal to that of men. (They may have quite different roles, but these are not superior/inferior hierarchies: they are equal in issues of interpersonal power and contribution to the tribe.) Therefore, in those societies women would have more of a say in their reproductive processes, when and how to have sex, when and how to use birth control, and so on. Certainly as women were increasingly empowered in the United States and Europe over the past 50 years, there has been a corresponding drop in population growth, whereas in those highly Catholic, Muslim, and Hindu countries where women hold low status and little power, overpopulation is rampant. Some see this as evidence that the very structure of a culture can affect its ability to control its population. In the Dane tribe of Indonesia, for example, most women choose not to have sex for five years after the birth of a baby. Among these people this system has worked for five thousand years to keep their population stable.

However they do it, though, tribal populations are stable in a way that reflects the available resources in their environment. Like healthy tissues in the body, they take what they need and nothing more. It works as if by magic, except it's a magic that also works for every other species of plant or animal in nature.

City-state populations, however, have a steadily increasing food supply because of their ongoing exploitation and conquest of surrounding lands. As a result, their populations grow without limit until they hit the sudden wall of famine or plague or the end of their energy resources. This has happened repeatedly ever since the first city-state in Mesopotamia experienced famine after they destroyed their environment through deforestation.

"But our nations are so stable. . . ."

Some readers may point to the countries of Europe that have stabilized their populations, such as Norway, Germany, and Italy, as exceptions to this rule. While it is true that these countries and others have succeeded in stopping the runaway growth of their populations (largely through contraception), they are still not living a sustainable existence.

These nations, like all other dominator Younger Culture city-states, are consuming vastly more resources than they produce. (Keep in mind that pumping or mining minerals or fossil fuels that are then consumed and destroyed is not "producing" anything.) Even though western European consumption of energy is lower than that of the United States per person, these nation-states exist in relative peace and prosperity only because they are able to exploit ancient sunlight that will one day be exhausted. In addition, they enjoy relative stability and prosperity because they can continue to convince governments in poorer parts of the world to let them exchange their goods with a local workforce in exchange for labor, and are allowed to extract the minerals and fuels from under the feet of the once-tribal people.

Although they may be stable at the moment, all city-state governments dependent on oil and/or growth are inherently unstable in the long run, because of the cultural stories they are based upon. Like a tumor or a Ponzi scheme, they depend on growth: when the GDP or GNP of these nation-states becomes negative or fuel supplies run low, they most often collapse into anarchy or lash out at their neighbors (as Germany did in the 1930s, and America and the U.K. did in 2003).

This is because of the centralization of those elements critical to life: food, energy, water, sanitation, and medicine. Their story is that centralization is good, that the wealthy and powerful are benign (at least so long as everything works), and that there will always be another distant land to supply cheap labor and natural resources (particularly oil).

Some countries, such as Norway, almost achieve stability. Life is good, literacy high, crime low, and poverty rare. But without pumping oil from the North Sea, Norway would face serious challenges.

Anarchy or tribalism?

If it sounds like I'm advocating dismantling the modern city-state, I am not. We've gone too far for that to be practical or to happen, and the experience with communism shows that when a dominator culture changes its economic or political system, what fills the void is merely a new form of domination. This book is not a call for revolution or anarchy.

Further, I'm also not suggesting that tribal life represents a utopian ideal. While many tribal peoples have leisurely and comfortable lives, there are also some who endure a difficult, brutish, and terror-filled existence. While no natural force in history has ever equaled the brutality, torture, and death visited upon tribal peoples by civilized members of city-state Younger Cultures (and certainly none before has ever annihilated them), nonetheless, many have lived difficult and painful lives at the whims of nature. Theirs is a sustainable life, yes—but not necessarily a comfortable one.

I'm not suggesting it has to be either/or, a return to tribalism or the destruction of what we call modern civilization.

Instead, we need to wake up to the cold, clear reality of the situation we've created in our world, and the reasons behind why it is the way it is: the dominator city-state Younger Culture that sees everything in the world as potential food or raw material for itself. If things don't change soon, it will grow and consume until there is nothing left to consume, and then our culture and our ecosystems will collapse, leaving billions of starving humans, polluted soil, air, and waters, and millions of dead species in its wake.

By adopting some of the lessons and worldviews of our ancestors—who lived in a stable fashion on the planet for at least 100,000 years—we can change direction and create a sustainable and livable future for at least a portion of the planet.

The least toxic form of city-state

How can we transform city-states—which are not going away any day soon—into egalitarian political and economic structures with tribal values?

Consider that the most important of all tribal values is that the aim of the culture is to provide safety and security for its members, and the job of each member is to make sure the culture can continue to provide safety and security to all. While this may be a new concept to many people who haven't read the works of thinkers like Marija Gimbutas, Riane Eisler, or Daniel Quinn, it's an idea that was explored in depth more than two hundred years ago by a group of people living on the eastern part of North America. They also looked askance at the seven-thousand-year-long history of domination and violence brought by their ancestors in what was called civilization, and decided there must be a better way for people to live.

To fully develop their ideas, they read about the Greek experiment on the island of Athens over two thousand years ago, and about the Saxons, who lived in what's now called England prior to the Norman invasion in 1066. These gave them clues, but to really develop their ideas they turned to the Iroquois Nation, inviting a group of Iroquois to their meetings, where they wrote their seminal documents.

One of those documents reads:

> We hold these truths to be self-evident, that all men are created equal, that they are endowed by their Creator with certain unalienable Rights, that among these are Life, Liberty and the pursuit of Happiness.—That to secure these rights, Governments are instituted among Men, deriving their just powers

from the consent of the governed,—That whenever any Form of Government becomes destructive of these ends, it is the Right of the People to alter or to abolish it, and to institute new Government, laying its foundation on such principles and organizing its powers in such form, as to them shall seem most likely to effect their Safety and Happiness. Prudence, indeed, will dictate that Governments long established should not be changed for light and transient causes; and accordingly all experience hath shewn, that mankind are more disposed to suffer, while evils are sufferable, than to right themselves by abolishing the forms to which they are accustomed. But when a long train of abuses and usurpations, pursuing invariably the same Object evinces a design to reduce them under absolute Despotism, it is their right, it is their duty, to throw off such Government, and to provide new Guards for their future security.

Certainly, the men who wrote this document were somewhat infected with Younger Culture stories. Some held slaves, and none thought women should be included in their new enterprise. Many, like Alexander Hamilton, still thought monarchy the best form of government, and others, like John Adams, referred to working people as "the rabble" and considered themselves a class apart.

But, looking past these obvious starting errors and challenges, for most the flame was lit. They'd lived in close proximity to Native Americans long enough to know that there was something important and vital in the Older Culture worldview, and that the single most important function of government wasn't to prevent people from wrongdoing or keep them under control, but, rather, to provide for their *life, liberty, and pursuit of happiness*. Those words were not chosen or written lightly.

And, for all the flaws and growing pains and even evils perpetrated by the young nation those men created, they had still succeeded in igniting an archetype—drawing deeply on tribal Older Culture values of egalitarianism—that is still, over two centuries later, despite the worst abuses and crimes of Younger Culture politicians, a beacon of hope and light to many in the world.

Democracy within a republic was the idea that Younger Culture city-states could be transformed to adhere to Older Culture values. Many within America, from her founding to today, rejected the Older Cultures so enthusiastically embraced by her Founders, instead binding themselves to fear, violence, police-state tactics, and a doctrine that the rich must be unimpeded in their efforts to acquire more without end. Others worked hard and valiantly to reform the nation so it could be a source of safety and comfort for all humans—and even all other forms of life. Over the centuries, the political pendulum has swung from side to side many times, and continues today.

The biggest challenge to Older Culture ideals within a city-state happened when a non-living, non-breathing entity rose up and claimed it should have the same rights and powers as the flesh-and-blood humans who had first written the United States Constitution.

The Robots Take Over

 [*The corporation*] *penetrating its every part of the Union, acting by command and in phalanx, may, in a critical moment, upset the government. I deem no government safe which is under the vassalage of any self-constituted authorities.*

—THOMAS JEFFERSON (1743–1826)

Imagine reading this in a history book:

About 120 years ago, spaceships discreetly appeared over cities around the United States and disgorged robotic machines to mingle among the humans and convert human life into sustenance for the robots themselves. While these machines are not alive and don't breathe, eat, reproduce, or die, in 1886 the United States Supreme Court—operating outside the boundaries of normal court procedures and without ever issuing a formal written opinion—altered the law of the land to recognize them as "humans" with all the rights of a human citizen. Actually, they gained more than all the rights of an American citizen, because the robots could live for hundreds of years, and if they broke laws or caused the death of real humans—even intentionally—they couldn't be put in jail or executed. (Technically it was possible that their operation could be terminated, but that had only happened a handful of times in over a hundred years and in each case they simply reinvented themselves.)

These robots began a systematic enslavement of humans and the Earth.

First, they approached the humans with a Faustian deal: give us your life and your loyalty, and we'll give you safety, security, and entertainment. At first, many of the humans refused, preferring instead to start their own family-owned or small, local businesses or operate family-owned farms. Over a 150-year period, the robots used their economic and political power to systematically destroy those businesses through a process the robots euphemistically called competition, and by the twenty-first century more than two-thirds of American workers depended on the robots or the government the robots had come to control for their security, health, and lifestyle.

The robots learned from watching the ancient Romans that "bread and circus" were essential to keeping people from becoming resentful of oppressors: so long as humans were well fed and entertained adequately, they'd submit to increasingly draconian intrusions into their private lives. Thus, the robots began to supply food and entertainment to the humans of the world.

The robots learned from watching the American and French revolutions that humans wanted to have leaders and governments who the humans believed were working in the interests of the humans. So long as humans could cast a deciding vote in their own government, they were mostly willing to overlook the robots' corruption of their elected officials.

Taking advantage of this, the robots proposed laws giving themselves control of the airwaves, and the elected officials agreed. The robots proposed laws giving themselves the right to produce cancer-causing wastes with the costs of cleanup paid for by the humans, and the elected officials agreed. The robots suggested that lands owned by the humans through their elected government should be rented to the robots for as little as five dollars an acre so they could extract from the lands all their mineral and oil wealth and keep the profits for themselves, and the elected officials agreed. The robots proposed laws that would give them the final say over the health and medical choices available to the humans, and the elected officials agreed.

As the robots grew in strength and power, immortal and unimaginably rich, they began to use the airwaves they now owned to offer their own servant-humans as elected officials, and the humans agreed. The robots had the power to control everything the humans saw, heard, and read: even the things the robots disagreed with but published, they'd manage to profit from and thus increase their power. Their servant-humans told the other humans that it was best to do things in the interest of the robots, and that the humans who were trying to protect the rights and lives of other humans were deluded, evil, "liberal." In the 1980s, enough humans came to believe this message that they allowed the robots to take over every aspect of the lives of the humans and of the government they had fashioned to protect them from "enemies foreign and domestic."

This wasn't the first time the robots had done this. In the 1500s, they took over several European countries, forcing the kings and queens of those lands to bow to their will and cede new lands to them. They started the first legal city in the new lands of America (Jamestown) and helped found the first legal state in the new lands of America (Virginia), a state that at that time stretched from the Atlantic Ocean to the Mississippi River. They regulated it and its citizens. The subservient government of England gave one of the robots (the East India Company) control of much of the trade in and out of the new American colonies, and when that robot proposed to the British parliament in 1773 that they eliminate all taxes on the robot's main product (tea) and give a tax refund to the robot, thus totally consolidating its power in the Americas and putting out of business the American small-businesses and entrepreneurs who were trying to compete with it, the Parliament complied.

That robot didn't yet own all the printing presses in the Americas, however, and word of this attempt to seize near-total control of the economic lives of the humans of the Americas spread: humans in Boston boarded one of the robot's ships and threw its tea into the Boston harbor. They started a war against the government the robot controlled, and even though the robot both funded and profited from supporting its British government, because it didn't yet control the American newspapers it lost the war. (The robots learned from this.)

The humans who declared war against the robots and their government in 1776 fashioned a new government that would provide exclusively for the needs of humans. They wrote that the purpose of their government was to protect and provide for the humans' rights to "life, liberty, and the pursuit of happiness" and that it served only humans ("government of, by, and for the people") and not robots. They passed laws that robots could only operate in single states, and that the lifetime of the robots was limited to 40 years. Additionally, at the end of each year the robots had to submit themselves to their state's government and prove they had been operating in "the [human] public interest" or be terminated. Many robots were terminated during the first century of the new American republic because they put their own interests above those of the humans.

The humans and their government passed laws that the robots couldn't create monopolies as they had in Britain—they were forbidden from controlling multiple aspects of a single business (for example, one could drill for oil but another would have to refine it; the refiner could make gasoline but couldn't sell it at retail), a restriction designed to prevent them from gaining enough wealth and power to represent a threat to the humans or their government. Another restriction for the same purpose forbade robots from owning stock in other robots. Their life spans and their business activities were strictly circumscribed, and they were prevented from interfering in the political activity that could lawfully be done only by voters/humans.

This situation seemed as if it would last forever and provide happiness, life, and liberty for the humans. It didn't.

The real robots

The non-living non-breathing entities that have risen up and taken over the world aren't generally called robots: they're referred to as corporations. They can live forever, don't fear death or prison, feel no pain, can change their citizenship in an afternoon, can tear off parts of themselves to create new entities, and don't need fresh air to breathe, healthy food to consume, or pure water to drink.

Prior to 1886, they were treated legally just like all other forms of human association (governments, churches, unions, guilds, small businesses, etc.) and had no *rights,* but only the *privileges* defined by "We, The People," the exclusive holders of human rights.

This was a critical issue, as one of the main points of the Founders was that *rights* should never again be held by an institution—like a church or a royal lineage—but should be held solely by individual human beings. Anytime humans get together to do something—be it running a business, starting a church, or even forming a government— that form of human association has only privileges, and "We, The People," who hold the rights, determine the specifics of those privileges.

In 1886, however, the reporter of the U.S. Supreme Court attached a specious headnote to the case of *Santa Clara County* v. *Southern Pacific Railroad,* saying that the Court had said that corporations should be considered "persons" and thus have access to the Bill of Rights, which the Founders had fought and died to provide to humans alone. Although this headnote openly contradicted what the ruling itself said, and had no legal standing or authority, Republican legislators (Rep. Roscoe Conklin and Sen. John A. Bingham) and corporations across the nation began to act as if the headnote *was* the ruling and demanded human rights for corporations.

They didn't want rights for unions, mind you. Unions to this day don't have the rights of persons. Nor did they want them for churches or small, unincorporated businesses or civic clubs. They didn't even want rights for governments. None of these groups today have rights. Just human beings . . . and corporations.

Writing in this specious headnote that corporations were entitled to human *rights* has brought about a dramatic transformation of American democracy away from the Older Culture vision of the Founders into a Younger Culture hierarchical system of rule-by-the-rich and the ownership of our politicians by corporations.

When corporations claimed the rights of humans, they threw off the legal constraints that had previously forced them to act transparently and in the service of the human community. One of the nation's largest and most conservative news organizations successfully argued in court that it had the right to force its reporters to lie in the evening news, and one of the world's largest athletic shoe manufacturers argued before the Supreme Court in April 2003 that it should have the right to lie in its public relations (and was backed up with *amicus* briefs filed by many of the nation's largest corporations).

In a landmark 1978 Supreme Court case, the Court struck down thousands of federal, state, and local laws that restricted corporate political activity when they ruled that the First Amendment of the Bill of Rights, which guarantees freedom of speech to human beings, also meant that corporations—which can't vote—could use their newfound "free speech rights" to contribute almost unlimited amounts of money to politicians, political parties, and advertising to support politicians they favor. It kicked off an era of the worst political corruption in the history of democracies around the world (most other democracies still limit corporations in their political activity or ban it outright), and could mean the end of American democracy as envisioned by the Founders.

Corporations claimed the Fourth Amendment right of privacy and succeeded in banning the Environmental Protection Agency (EPA) and the Occupational Safety and Health Agency (OSHA) from performing surprise inspections, kept secret information about their political activities, and even hide evidence of toxic or dangerous products and outright crimes. (The Fourth Amendment doesn't seem to refer to corporations when it says "The right of the people to be secure in their persons, houses, papers, and effects, against unreasonable search and seizure shall not be violated.")

Corporations claimed the Fourteenth Amendment—originally passed after the Civil War to free the slaves—and argued that when communities tried to keep out predatory corporations, it was a form of illegal discrimination the same as if a restaurant chose to serve people with one color of skin but refused to serve people with a different color of skin. (The Fourteenth Amendment is similarly clearly written to apply only to humans.)

Corporations are non-living, non-breathing entities that can act on the world, just like robots. They're a legal fiction. They can feel no pain. They can live forever. They can't be put in prison. Buckminster Fuller said, "Corporations are neither physical nor metaphysical phenomena. They are socioeconomic ploys—legally enacted game-playing—agreed upon only between overwhelmingly powerful socio-economic individuals and by them imposed upon human society and its all unwitting members."[1] Prior to 1886, they were referred to in U.S. law as "artificial persons," similar to the way science-fiction authors portrayed robots.

Corporations today play a role that was historically filled by royal courts: they control most of the wealth and exert power over the lives of most citizens. Their CEO/kings are unapproachable and live lives of mind-boggling luxury. They've become the rudder that is steering the ship of our culture, and they're steering it by their prime value—growth at any expense.

In biological systems, when one part of a complex organism rises up and decides to take energy and resources from all others because its sole focus is growth, we call it cancer. Many suggest that modern transnational corporations have become cancers on the body of democracy, destroying social and political stability, and are endangering life on Earth, including human life.

Citizen groups across the world are working to minimize the damage being done by these robotic entities, ranging from "abolish corporate personhood" campaigns (which I wrote about in my book *Unequal Protection: The Rise of Corporate Dominance and the Theft of Human Rights*) to attempts to roll back treaties and laws that have handed such awesome wealth and power to corporations.

The fate of democracy—and, therefore, the fate of the remaining Older Culture parts of what we call modern civilization—hangs in the balance.

But What About Darwin?
Isn't the Victor Right?

 The provision of the Constitution giving the war-making power to Congress was dictated, as I understand it, by the following reasons. Kings had always been involving and impoverishing their people in wars, pretending generally, if not always, that the good of the people was the object. This, our Convention understood to be the most oppressive of all Kingly oppressions; and they resolved to so frame the Constitution that no one man should hold the power of bringing this oppression upon us.

—ABRAHAM LINCOLN (1809–1865),
February 15, 1848

The meeting with [Iran's foreign minister before the Gulf war] permitted us to achieve congressional support for something that the President was determined to do in any event.... I think we would have gone ahead anyway ... even if we had lost the vote.... I don't acknowledge that I have to have congressional approval.

—JAMES BAKER, former secretary of state,
speaking about the start of the Gulf war, 1991

U.S. NOT INTERESTED IN IRAQ'S OIL

—*PETROLEUM WORLD* headline
October 31, 2002

One argument that people (particularly conservative talk-show hosts) put forward when the ideas in this book are presented is, If the tribal way of life was so good, how come we conquered them? Doesn't "winner" mean "superior"? The easy answer to that is, Was Hitler's way of life superior to that of the French and the Poles?

The classic history of America's occupation by Europeans has it that we either found huge areas of "unused" land that the ignorant savages had never figured out how to use, or else we "conquered" them because we were smarter and more civilized and therefore had technologies like guns, which guaranteed we'd win.

In fact, neither of these views is true, as can be found in dozens of early histories of this country. There were at least two attempts to conquer or colonize the "savage lands" of North America that were successfully repulsed before the Puritans set up house in 1620. Even with superior force of arms, Europeans didn't have the survival skills to successfully live in competition with the Native Americans. Instead, as with the Inca, it was disease that made possible our colonization of North America, a fact that's oddly overlooked in most high school history books, albeit well documented in college and other texts, which don't have to be vetted by the State of Texas in order to be sold to schools.

Europeans, like the "severely pox-marked" George Washington, had suffered from smallpox for centuries, and after the epidemics, those who survived had strong genetic resistance to the disease. While it and others such as chicken pox, influenza, bubonic plague, and hepatitis were common among Europeans, they caused death in a relatively small number of cases.

Not so among the Native Americans. Wherever Europeans went, Native Americans died by the hundreds of thousands to the millions. (Estimates cited by William McNeill put the Native population of the Americas at 100 million people when Europeans first began aggressive colonization in the early 1600s. Today there are fewer than one million pureblood ancestors of these people.)

For some years before the Puritans landed in Massachusetts, the natives had been trading with Dutch, French, and British fishermen and itinerant coastal traders. They spread plague among the natives—most likely smallpox—so completely (but accidentally) that by the time the Puritans showed up in 1620, Robert Cushman, a British eyewitness, said that fewer than 5 percent of the Native Americans were left alive. Entire villages were wiped out, the ground covered with skulls and bones, the few survivors in most cases having fled westward . . . carrying the disease with them.

That between 90 and 95 percent of the Native inhabitants of New England were wiped out by disease was seen by John Winthrop, the then-governor of Massachusetts Bay Colony, as a "miraculous" sign from God. He wrote a letter to a friend in England, in 1634, saying: "But for the natives in these parts, God hath so pursued them, as for 300 miles space the greatest part of them are swept away by the smallpox which still continues among them. So as God hath thereby cleared our title to this place, those who remain in these parts, being in all not 50 [Native American persons], have put themselves under our protection."

As conflicts between Europeans and Natives began again over the following decade when the Europeans began pushing west, the Puritan minister Increase Mather wrote: "God ended the controversy by sending the smallpox amongst the Indians. Whole towns of them were swept away, in some of them not so much as one Soul escaping the Destruction." Ultimately, plagues followed Native Americans from Florida to Maine, Massachusetts to California, "preparing the way" for the colonization of America by Europeans. As Puritan author of *On Plymouth Plantation* William Bradford wrote in 1632 of a problematic local Indian village, "It pleased God to afflict these Indians with such a deadly sickness, that out of 1,000, over 950 of them died, and many of them lay rotting above ground for want of burial."

And Charles Darwin wrote in 1839, "Wherever the European had trod, death seems to pursue the aboriginal."

Nonetheless, the Darwinian "survival of the fittest" view (while conveniently overlooking plagues) has become a central part of the stories we tell ourselves about how the world works. But survival in the recent past does not ensure survival in the future. To realize this, all you have to do is look at the thousands of species that survived

throughout history only to be exterminated in this century. To predict what's needed for the future, we can't consider just the past—we
must look into the future.

The highest value of tribal societies is cooperation. They practiced it within their own tribes every day. They demonstrated it to
Europeans when we first arrived on these shores and they helped us
plant crops and survive the initial winters.

The Iroquois cooperated with Ben Franklin when they allowed
him to attend their tribal meetings, where he learned about their
thousand-year-old Great Binding Law of the Iroquois Confederacy,
which existed before Columbus arrived and governs the Iroquois
Nation to this day. Franklin took their idea of a governmental system
with internal checks and balances, a separation of the judicial and
legislative, and elected representatives, and shared it with James
Madison and Thomas Jefferson. The three then integrated these ideas
into the Constitution of the United States of America. Franklin, Jefferson, and Madison all wrote of this extensively in their papers, and
Franklin invited 42 members of the Iroquois Confederacy to attend
the Albany Plan of Union in 1754 when the first try at a representative democracy was discussed. Franklin later said in a speech to the
Albany Congress: "It would be a strange thing . . . if six nations of
ignorant savages should be capable of forming such a union and be
able to execute it in such a manner that it has subsisted for ages and
appears indissoluble, and yet that a like union should be impractical
for ten or a dozen English colonies."

The early colonists decided, however, that they knew better than
the Iroquois how to form a government. While they emulated the
bicameral legislature, Supreme Court, and clearly defined limits on
the power of the central government that the Iroquois had had in
place for thousands of years, the colonists had a lingering affection
for the monarchy. George Washington, who argued unsuccessfully
that as president he should be addressed as "his highness," was among
those pushing to add a chief executive, or surrogate king, to our system of government.

Nearly all the colonists agreed that the Iroquois system of the
(mostly male) elected representatives being elected *only* by the tribe's

226 THE LAST HOURS OF ANCIENT SUNLIGHT

women (who alone had the power to remove them from office) was a mistake. The colonists altered this so that only men could make such a decision.

They also decided to ignore the Iroquois rule, which persists to this day, that *all* decisions of "importance" (such as waging war, changing national boundaries, altering relationships with other tribes, etc.) must be submitted to the local electorate by the elected representatives for discussion, debate, and decision. Instead, they created the system we now have where such decisions are made daily without consulting the electorate.

In contrast, the core value of our culture is not cooperation but power. Power of gods over men. Power of one group of men over another; power of men over women; power over property (who owns what, and who's not allowed to have it). Power of humans over the natural world. *Power.*

It's hardly surprising, then, that a culture that values power over all else would wipe out a culture that holds cooperation as its highest value. But does that mean that the culture that values power is better? Or that it will survive forever? Or even that it will survive one-hundredth as long as older cooperative cultures have?

Perhaps the neo-Darwinists are right, and those civilizations that survive are those that are superior. But the battle for survival is a saga that is not yet over, and preliminary findings indicate that power-centered cultures have eventually always self-destructed.

Remember that after tens of thousands of years of relatively peaceful co-existence, a small group of Mesopotamian people rose up and decided to exalt power and domination above all other people and things. They won their wars and grew in numbers, and believed that their expanding powers proved their superiority, correctness, and blessings of their civilization.

But then they collapsed.

From the ashes of their funeral fires, as their famines and plagues and despoiled lands were fading into dim memory, another group gave it a try, and another civilization emerged. It, too, collapsed. And another, and another: the Mesopotamians, the Greeks, the Romans, the Huns, the Ottomans, the Inca, the Aztecs.

Will today's incarnation of the cultural, political, and economic system that wields power as its primary tool also collapse? Will the tribal people be the only ones left?

Is it possible that the meek shall indeed inherit the Earth?

If the signs we see all around us are accurate, it may well be that our neo-Darwinists are right . . . but that they picked the wrong culture as the "superior" one, at least in terms of its ability to survive over the long haul.

PART III

WHAT CAN WE DO ABOUT IT?

All humanity once knew how to live in concert with nature, how to live sustainably, and some humans still do know how. But after more than five thousand generations of cooperation, domination crept in, and it spread around the world like influenza, infecting the whole planet within fifty generations.

Modern civilizations have come to believe that the paths to a better life are consumerism and the manipulation of the "machine of nature" to our advantage. Despite overwhelming evidence to the contrary, these twin tenets of Younger Culture are still regarded as our salvation.

For example, in October 1997 the U.S. timber industry, along with Newt Gingrich, announced that opening more federal forestland (particularly in Alaska) to commercial logging would actually "help the problem with carbon emissions" because the trees would be cycled into paper and houses, thus "stabilizing" the carbon. Apparently, they hadn't thought ahead to the next year: paper and houses don't breathe in carbon dioxide and exhale oxygen, don't make topsoil, and don't stabilize land or the water cycle. On the altar of short-term profit, before the false god of consumerism, we are plundering the world, putting our children's future at risk, and even most educated people don't realize how or why it's happening.

Even in the face of the accelerating damage we are doing to the planet and our own species (through pollution of our environment, among other things) there is the possibility of change. The David of a new way of life stands before the Goliath of politicians and corporations, and the small stone of ancient stories from Older Cultures may well strike the Younger Culture's forehead and then ripple out throughout the world in ways that will give birth to a new world.

Much of this book has been devoted to how bad things are, how disastrous they could become (although it is optimistic compared to some), and how we have arrived at this crisis in the history of the world and the human race. It's been a large part of the book, because it's really a story that spans five to ten thousand years. In order to fix the future, we must understand the past.

Now to the future. The answers are in so many ways straightforward once we see through the lies and distortions of the past, and can learn to ignore the constant drumbeat of a culture devoted to domination and exploitation. There are specific things we all can do. Most are small and simple, and have to do with how we think and see and hear and feel. Some are larger and more dynamic. All begin with one person understanding how things are, how they got this way, and that there are alternatives. Right now, that person is you, and then you can pass this understanding along to others, and they to others, and on and on.

Part III shows bright hope for a warm and positive future. You'll learn specific tools and techniques, ways you can change your world and the world around you. The topics are grouped into these categories:

Transform ourselves

- There is a "morphic field" wherein we are all connected, identified by Rupert Sheldrake and referred to by Carl Jung as the "collective unconscious," where, as we each individually begin to change our way of thinking and living, our actions echo out into the larger world.
- In this way, we can find new "stories" we can begin to use, to change how we think about what happens in life.
- The most important part of personal transformation leading to planetary transformation is to become fully

alive, alert, and aware of our surroundings and the divinity everywhere.

Change our technologies

- We can begin to use our remaining oil to help us develop the next energy solution.
- Both for planetary transformation and for survival through possible tough times, we can learn to become independent of the power utilities and other huge corporations.
- Conservation is something we can all begin now (often not in the conventional sense), and this will slow the rate of planetary deterioration.

Change how we think of and use science

- The *best news* is that science, which seems to have been one of the players in the destruction of the planet, is now showing and telling us how everything literally is interconnected and conscious.
- Science shows that our thoughts and even our smallest actions do make a difference.

We can learn many of these lessons simply by reconnecting to the wisdom of our ancestors

- They lived a life of "spiritual ecology" from a view of the sacred nature of *all* creation.
- They taught and still teach specific ways we can wake up to *life.*

We can build communities that work

- Thousands are forming a new generation of "tribes"—small intentional communities where people care for each other and live sustainably.
- These communities are lights in the transformation of the planet, as well as places where an often extraordinary quality of life is found and lived.

- You can connect with or create such a community once you understand how they work.

Transforming culture by transforming politics

- Politics is the concrete expression of a cultural worldview, and the Founders of the United States drew heavily on Older Culture perspectives in creating this nation.
- By awakening to the agendas of the various political groups, and participating in political activity, we can reintroduce Jeffersonian Older Culture perspectives into the body politic of America and the world.

Each of us is descended from humans who lived in small groups, cared for each other, and met their needs in sustainable ways. Democracy is in our genes, and cooperation is our history. We have much to learn or, more accurately, much to remember.

The New Science

 Science without religion is lame. Religion without science is blind.

—ALBERT EINSTEIN (1879–1955),
Ideas and Opinions

We live lives that are very much the product of science. To reject it in whole, to think that we can just walk away from it all and instantly return to a tribal life as it was practiced hundreds and thousands of years ago is a fantasy. It isn't possible, and probably wouldn't even be desirable. There are some—many—benefits that technology can offer us. What is necessary, instead, is to begin the process of putting science into perspective.

How are we to view the world, or even the entire universe?

Our Younger Culture has used the reductionist, atomistic perspective of Aristotle, Newton, Descartes, and others to suggest that the world is simply a machine. It's made up of a *lot* of interlocking pieces, to be sure, but a machine, nonetheless. Each part can be reduced to its individual pieces, this story says, and if they are broken, they can be repaired.

When my car was "broken" in an accident last week, we took it to the local garage to have some of its parts replaced. It was taken apart and put back together again, and today I will pick it up from that garage with every expectation that it will run just as it had before with no noticeable difference. It may even run better, as they will look it over for things that need to be adjusted or tuned or replaced.

But is this really how the natural world is?

When we look at things in the world, we find not machines but living things. Trees, flowers, insects, birds, mammals, humans. Like many students of modern medicine, I once believed that all were machine-like, and that, like that car, everything could run again when put back together.

When I was fourteen, I studied biochemistry during the summer term at Michigan State University. My lab-mate and I decided to undertake an ambitious project: to kill a cell and bring it back to life. We chose an aquatic plant that had single cells so large you could see their nuclei, and extracted the nuclear material from several cells. Then we injected into a living cell's nucleus a compound that would dissolve that cell's DNA into free nucleic acids. We inserted a second compound that neutralized the first, and then attempted to inject into that same nucleus the DNA we'd extracted from other cells. Our experiment succeeded only insofar as it taught us that a dead cell cannot be brought back to life.

Dr. Frankenstein's creation notwithstanding, there is a significant difference between machines and living things. Both share a complexity that makes them greater than the sum of their parts. A pile of parts and an assembled automobile are quite different things, even though the parts may be the same. The difference is the organization of the parts, the system or structure that is imposed on them.

Living things, similarly, have a structure that is intrinsic to their uniqueness. All the disassembled parts of a cow, for example, will not moo and walk around. Prior to their disassembly, they were organized in a particular fashion that gave them cow-ness.

The difference between a living thing and a machine is not structural. A machine can be stopped, taken apart, reassembled, and then be the same machine. This is not true of a plant or animal. Living things, once stopped in their life processes, cannot be revived.

There are those who would argue that this is simply because we haven't yet figured out how to restart life. The cryogenics movement, for example, is based on this simple article of faith: that someday we will know how to reactivate reassembled or once-stopped life, but there is no evidence whatsoever to support it.

What evidence there is shows that there is something fundamentally different between a person and a corpse. This difference is worlds apart from, say, a car that is running and one that is sitting at the curb turned off. This is because a machine is organized on a particular system or matrix. A plant or animal, on the other hand, is organized in an utterly mysterious fashion, which we may never fully understand, and when it becomes dead, something beyond just its organization has "departed" from it.

That we don't understand how life-forms are organized is a fact that many scientists would prefer to ignore. Modern medicine attempts to reduce the human body and mind to machine status, and repeatedly discovers that these are far more complex than previously imagined. The poorly understood interplay between body and mind, as just one example, has confounded physicians since the time of Hippocrates.

So we have this difference between machines and life-forms: the former can be taken apart and put back together, whereas the latter are vibrating with some unknown essence we call life that vanishes forever and cannot be restored.

This brings us back to how we should view the world and the universe it is set in. When we look at the natural world around us, do we see machines? Are the trees and plants inanimate structures of mineral and energy? Are animals mere collections of organs and parts? Are the delicate life-systems of the oceans and land and atmosphere something that can be stopped and then restarted by simply throwing in the requisite chemicals and amino acids?

The first person's view

Living out in the country brings some interesting insights. Last year I met a Native American medicine woman. She said that when she went into the forest or fields, she didn't see just trees and plants and animals, but saw their spirits and heard and felt their consciousness. The trees told her of their lives, their pain and joy, the plants told her which could heal and which could harm humans. The animals gave her instructions on how to live in harmony with the land, and the land spoke to her in an identifiably female voice.

"This is how native people have seen the living things on this land for eternity," she said. "You whites were blind when you arrived here, and are still blind."

At first, I applied the "white European" atomistic view of the world to her words. She was anthropomorphizing, projecting her own thoughts and desires onto other living things. She wasn't really hearing them but was only speaking in metaphors, although she had insisted she was speaking literally; she was misinterpreting natural phenomena, thinking, for example, that the tree was nodding or gesturing to her when, in fact, it was simply the wind, mistaking a bird's natural territoriality for an attempt at personal communication.

Then I realized that I was doing exactly what I'd assumed she was doing: projecting my own view of things onto her statement. When Western science looks at an ancient people and begins this type of analysis, it is every bit as much a projection, an article of faith, the reflection of a belief system, as we are asserting when we are appraising the other.

So I went out into the forest around our home in Vermont.

"Is there conscious life in you?" I said softly, looking at the maple and spruce. They gently waved in the wind, a distant bird began to sing, and I could smell the fresh scent of moist earth.

I wondered if the entire forest might answer me with, "We are alive," but instead I got a powerful sense of individual aliveness from each life-form I looked at. Each tree, the bird and the chipmunk, the soil under my feet teeming with microorganisms; each seemed to assert its own individual aliveness. Like the individual musicians in a symphony orchestra, they played together to create a beautiful sound.

I raised my hands, palms out, imagined my life co-mingling with that of the forest around me, and was filled with a thrill at touching the life of the Earth.

This is a different type of science—the science of the first people who viewed the life of the planet. When Jack Forbes, who first wrote about *wétiko,* told me, "Native people do not necessarily believe that only humans can talk," I felt an immediate contact with an ancient knowledge, something that has been hidden and lost in our attempt to make everything in the world fit into our machine-like worldview. Just as our Younger Culture at one time couldn't imagine the Earth was round because the concept didn't fit into our reality, we have also rejected much ancient knowledge that has value because it didn't fit in with our Cartesian worldview.

Try it yourself. After you set down this book, walk out into the natural world and try communicating with—sensing and speaking to—the plants and animals around you. Find within yourself the place where you sense the presence of life, and from which you can reach out to other life, to all life. From this place of seeing *all* life as *sacred* life, you can then begin to thoughtfully consider other things you can do to create a sustainable future.

Physics discovers consciousness

Physicists point out that physics was the first discipline of what is called modern science. When medicine was still filled with concepts of spirits invading bodies, and astronomy was indistinguishable from astrology, Aristotle laid the foundations for modern physics as he began an inquiry into the ultimate nature of reality. Things were made of smaller things that were made of smaller things . . . until you got to the smallest thing that, Aristotle thought, is the atom.

Physics has always led the other sciences, because every other science deals with some aspect of "reality" and physics concerns itself with what—at its very core—that *reality* is. Chemistry without physics is inconceivable; biology without chemistry is incomprehensible; medicine or genetics or farming are unimaginable without biology. Every science ultimately builds on the foundation of physics. Similarly, the core of the scientific model and scientific methodology grew out of the study of physics.

Until now. Today, it seems, the other sciences are either huffing and puffing to catch up with physics, or else trembling at the implications of physics' most recent discoveries. The good news for us today is that the discoveries of physics show that we are much more connected to the rest of the universe than our culture has been telling us. Science has, in a real sense, just recently caught up with what the Older Cultures had taught our ancestors all along.

Consider, for example, the simple electron. When first discovered, the electron was thought to be a tiny particle, flying around the nucleus of the atom, which itself was made of protons and neutrons. The orbits of electrons around atomic nuclei seemed to be organized into shells, and the model most often invoked was that of the solar system with electron-planets orbiting the nucleus-sun.

Proof of this was found in 1906 when Lee DeForest learned how to heat a wire (a "cathode") to produce a cloud of electrons, and then use a positive electrical charge to form them into a beam (a ray) and direct them at something. The flickering image on the screen of your television (invented later, building on DeForest's work) is created by a stream of electron particles hitting phosphor atoms on the inside face of the screen, causing them to glow. It's a CRT: a cathode ray tube.

One day scientists tried sending a stream of electron particles against a metal panel with two slits cut into it, placed in front of a phosphor-coated piece of glass. What happened shocked them, and turned the world of physics on its head. If the electrons were particles, the beam should have scattered through the two slits and formed two neat little slits of electron-impact spots on the phosphorus, just as if they were fast-moving grains of sand. Instead, though, the electrons turned from particles into waves, flowed through the slits like light or sound would, and produced a pattern of overlapping ripples, just as if two pebbles had been dropped in a puddle.

"This is impossible!" the scientific world shouted . . . until the study was replicated dozens of times in dozens of different ways.

Even more amazing, some of the follow-up studies showed that when the electrons could "choose" to behave like a wave or a particle, they always chose to be waves . . . unless somebody was watching, in which case they snapped into being particles. Without observers, electrons (and everything else, it turns out) exist only as a mathematical *possibility*, a potential, just as a roll of movie film at your local theater is a "potential movie-reality." Only when somebody watches—a living thing observes—do the electrons climb out of their film can and present themselves on the movie screen of our reality-world as particles.

In a way, it's like the story of King Midas, who chose as his wish that everything he touched would turn to gold. In a similar fashion, many physicists now believe that everything we look at turns to reality as it is seen (although, unlike with Midas, our "reality" apparently dissolves back into probability once we look away). In one of the best books on this topic for non-scientists, *The Holographic Universe*, author Michael Talbot quotes physicist Nick Herbert as saying his learning of this has caused him to think of everything behind him as "a radically ambiguous and ceaselessly flowing quantum soup" that, when he spins around and looks, snaps into physical reality so fast as to appear seamless.[1]

But where does this "soup" come from, and what is it made of?

In another experiment, physicists discovered that if they split a subatomic particle into two pieces, those two half-particles went flying off into space, each spinning like a baseball in a direction opposite of the other. However, when physicists put one of the particles through a slit that would change its direction of spin, they learned that the other twin particle—miles away by that time— would *instantaneously* change its spin to correspond with that of its modified twin. The experiment was carefully and cleverly designed to eliminate any chance of communication between the two particles.

Again, the scientists were aghast. The second particle didn't change its spin after a long enough time for the information about the direction of spin to reach it at the speed of light—it changed its spin *faster* than the speed of light: instantaneously.

The implications were mind-boggling. If, for example, you wanted to talk with a person on a star five million light-years away, and you tried communicating using a beam of light, from the time you blinked your signal until the time that person could see, it would be five million years. Such communication hardly seems practical, given the normal human life span. Even with a star only 50 light-years away (among the closest), it would be extremely cumbersome.

But if there was a star somewhere in between, and that star was blowing out spinning particles (as many stars, particularly neutron stars, do), we could communicate *instantaneously*—as if we were using the telephone across town (faster, actually, because the phone still has to use electrons that travel just slightly below the speed of light). Theoretically, all we'd have to do is modulate (change the spin of) a stream of particles formed from in-star particle-splits, and the person on the other side of that vast stretch of space could see the changes in their particles instantly.

Of course, at first it seemed that this must be impossible. One of Einstein's foundational principles—the Gregorian Chant of physics—was that nothing can exceed the speed of light. It takes five million years for your message sent at the speed of light to travel the distance light would travel in five million years. In 1935, Albert Einstein joined with two colleagues in publishing a paper that pointed out that while the evidence shows that *something* is *apparently* traveling faster than the speed of light, it's still impossible according to the math. It's a paradox. And so it was (and is) referred to as the Einstein-Podolsky-Rosen (EPR) Paradox.

Danish physicist Niels Bohr, however, pointed out that Einstein, Podolsky, and Rosen were making a fundamental error in their assumptions about the particles being studied. They were assuming that the particles were *things,* that they were each separate from the other, and each had an independent existence. What if, Bohr asked, the two particles—even though separated by millions of miles—were instead part of the *same* thing, still two components of the original split particle, and had no separation so far as they knew? And, therefore, since they were both part of a whole, when one was affected the other would be similarly affected *at the same time?*

As repeated experiments proved that Bohr was probably right, his interpretation of Einstein's math and commentary came to be known as the Copenhagen Interpretation and the phenomena he described were called *non-local phenomena* or *non-locality*. It's now considered by many to be a fundamental principle in quantum physics, even though it implies that time and space are very different from the way we previously thought of them. They're more of an idea in some universal mind than physical reality in some universal reality.

More recently, British scientist and author Rupert Sheldrake pointed out how animals often seem to behave in a non-local fashion. When a certain number of birds in England learned how to open the tops of milk bottles left out by milkmen in the 1930s, birds all over Europe suddenly began to do it. The speed of the transmission of the behavior defied any possibility that one bird had traveled to another and taught it . . . and the English Channel added a further barrier, as these were *not* migrating birds.

Consciousness, the new physics implies, brings the universe into existence, and consciousness is not confined to any one place. One interpretation is that *the universe is made of consciousness . . . and nothing else.*

This phenomenon of instantly shared remote knowledge, which Sheldrake calls *morphic resonance,* implies that humans can behave in a way analogous to Einstein and Bohr's subatomic particles. When enough people learn something new, suddenly there's a snap or shift, a resonance in the human morphic field, and *everybody* is awake to the new information. There are countless examples of this, from the speed with which jokes travel around the country to how cultures shift and change without apparent organization.

You *do* change the world every day

Thirty years ago, I spent a few days with a renegade Sufi teacher in San Francisco. He described his notion of reincarnation, which I think is an interesting metaphorical analogy to how morphic resonance and non-locality imply that we're all constantly changing the world.

When we die, he said, our consciousness dissolves into what he called "the cosmic soup." All our thoughts, dreams, fears, experiences, and everything—it all goes into the soup-pot, forming "a huge cosmic goulash, with everybody mixed together with everybody else." When a new baby is born, he said, "the cosmic cook" picked up his ladle, reached into the cosmic soup-pot, and drew out enough of the soup to fill a human body/soul. This was poured into the new human.

It was an interesting concept, and I have no strong opinion one way or another on its validity. However, I particularly like the meaning he drew from it. "Because we all come from the same soup," he said, "we all have an obligation to make the soup happier, lighter, better tasting. Every thought we think and every action we take will eventually become the soup, and so be poured into one of our descendants. So our actions, our thoughts, our words—even the most seemingly insignificant—are important."

Looking at Einstein's, Bohr's, and Sheldrake's work, however, the question arises, Why wait until we die to add to the soup? In fact, all the available evidence, from physics to psychology to common sense, tells us that our actions now, today, *this moment as you read this book,* are influencing everything and everybody in creation.

Practice small acts of anonymous mercy

So where do we begin?

In the Sermon on the Mount, Jesus pointed out that when we do "good works," we should do them without other people knowing that we did them. This is a difficult task: you have to continually keep an eye out for such opportunities.

Many people, looking at the enormity of all the problems facing the world, feel depressed, overwhelmed, and apathetic. They often give up.

But there is great spiritual and cultural power in performing small acts of mercy. They echo farther than most people realize, and begin a "morphic resonance" process of putting out into the air—in a way that becomes culturally contagious—the millions of small steps that must be taken worldwide to save our planet and our species.

On some level, we are all connected. When you save the life of another living being—even a worm or a weed—you are putting into the air the saving of lives. Small acts of mercy are among the most transformational spiritual activities a person can engage in, which is probably why Jesus and those teachers and prophets before him repeatedly put such emphasis on them.

A Cree Native American storyteller and teacher told me: "According to my tradition, from the beginning of creation, every morning, when the sun comes up, we are each given four tasks by our Creator for that day. First, I must learn at least one meaningful thing today. Second, I must teach at least one meaningful thing to another person. Third, I must do something for some other person, and it will be best if that person does not even realize that I have done something for them. And, fourth, I must treat all living things with respect. This spreads these things throughout the world."

For example, in most of the world's Salem Children's Villages (communities for abused children around the world, first started by Gottfried Müller in 1957) there are stables with horses for horseback riding. I'd known for years about the horses in Stadtsteinach, Salem's German headquarters: I had seen them perform dressage, had fed them, had walked to their stable and given them apples every evening with Gottfried Müller, my mentor, after dinner in the Salem guest house. What I didn't know at first was where the horses came from.

Over time the story came out, since Herr Müller doesn't often talk about the "good deeds" he does. He'd been in a train station and a train came through carrying horses from Czechoslovakia for the sausage factories of Germany. Seeing the horses, he inquired if it was possible to "save" any of them. The sausage company agreed to sell him a few, and those horses became the original horse population at Salem.

I'd often wondered why the Salem horses seemed to exert such a powerful attraction to both the children at Salem and visitors. Now I believe it may have to do with Gottfried Müller's quiet action in saving their lives.

In October of 1997 I was in Stadtsteinach with Herr Müller over breakfast. A staunch "independent Christian" (he will join no organized religion) but fond of Christian and Jewish metaphors, he said, "You know, in the balance scale of good and evil, there is much power and weight on the side of pain and torment and evil in the world. The story of Job tells how many different powers evil has, to create wars, to make pain, to afflict people, even to create what look like miracles. But there is one ability that Satan does not have. It is an ability that only we have. And, because he does not have this ability, even when we use it in very small ways, it is a great weight for good on the balance scale of the world."

"And what is this ability?" I said.

"Barmherzigkeit," he said. This is a German word that means small acts of mercy, performed with compassion. "And, as Jesus said

in the Sermon on the Mount about the widow who gave a penny, it is often the smallest, most anonymous acts that create the loudest thunder in the spiritual world."

Your actions, words, and even your thoughts have a powerful spiritual and real-world effect, whether others know about them or not. We are each like miniature transmitters, putting out into the air whatever we're about at the moment. This is why monasteries and retreat centers and the Salem communities around the world are so important: they're spiritual beacons, and they radiate the light that they're producing into the non-locality, the morphic field of the real world.

No matter how overwhelming the problems of the world may seem, you *do* have an effect, even if nobody ever knows what you've done. For example, prayer has been demonstrated in double-blind, scientifically controlled experiments run at Harvard University to speed healing, even when the people praying and the people healing don't know each other, have never met, and are located in different parts of the world.

And, a step beyond prayer, consider how powerfully you could help transform the world if you were to connect directly with the source of the field of all reality. . . .

Reconnect with God . . . directly

Most of the world's major religions sprang from tribal cores, and those tribes had a morphic field of their own. Jews still speak of the ancient "twelve tribes," and Jesus said many things that were antithetical to the Roman Empire city-states of his day but make perfect sense if one views people of his time as choosing to live either tribally or in city-states. They make even more sense when viewed through the lens of quantum physics. Similar roots and teachings can be easily found in Hinduism and Buddhism.

But the sects that have arisen from all of these religions have been contaminated and taken over by the hierarchical power structures of city-states and their dominator mind-sets. Their essential and original truths have been lost.

Gottfried Müller once said to me, "We need a new Christianity, because so many of the churches are lost. They no longer teach what Jesus said, but only *about* what he said."

As any mystic or person who's had a true religious experience can tell you, there is a core of truth, of power and beauty and love, in our religious traditions: a point of non-locality, what the Buddhists call beyond the beyond. It has been experienced numerous times in the past, and some of those who experienced it have left records of it. Our consciousness is part of—or capable of connecting with—a larger Consciousness that pervades all life and all creation.

This intelligence did not just bring forth all of creation: I believe it is all of creation. As such, it is beyond names, beyond petty considerations of ritual and dogma, beyond preferences. It is. "I am that I am," it said to Moses. "Be still and know Me."

Millions of people have cut through the hierarchy and directly touched this unnamable God, directly experienced the intelligence of the universe, felt the power as love, and seen the world with new eyes as a result.

The early Christians took what Jesus said literally, most likely because they had the direct experience of God that he described. "You are the sons of God." "These things I have done, you will do also." "The kingdom of heaven is within you." "Do not worry about tomorrow." "Forgive the past." "Pray in secret instead of in public." "Don't let anybody know about your good deeds." "Do not acquire wealth, and if you have done so, give it away." "Don't gather food into barns." "Give to anybody who asks you."

Legend has it that nearly all Jesus' original disciples suffered agonizing deaths—being boiled in oil, crucified upside-down, having their skin ripped from their bodies, being fed to lions—because they had adopted an Older Culture tribal way of life. They were enemies of the governments of their time, in part because they refused to participate in the exploitative systems: they shared all that they owned, had little money or medium of exchange, and tried desperately to get the dominator world then ruled by the city-state of Rome to change its ways. They never succumbed to the dominator mental illness,

never took up swords and spears to battle their heathen Roman enemies. Nor did they ever attempt the frantic climb up the consumer
and political hierarchy in the culture surrounding them. Jesus challenged the hierarchical, dominator mentality of one person being
superior to another, when he said, "But be ye not called Rabbi
[priest]," "Call no man your 'father' upon the earth," "Neither be ye
called masters," and, "He that is greatest among you shall be your ser-
vant. Whosoever shall exalt himself shall be abased; and he that shall
humble himself shall be exalted."

* * *

So we can now see that science is proving the existence of something
it once thought disproved: the living nature of the universe and the
interconnectedness of all things. That in stepping back from the
intrusions and distractions of our corporate-driven culture, and in
reaching out to the divinity both within ourselves and within nature,
we can find a power and purpose and deep meaning to life. From this
place, from this new vantage point, we can see the essential insanity
of the *wétiko* dominator lifestyle, and when enough people figure this
out, we will turn around on the destructive road humanity is now
following.

But how many people need to know this?

A flyer I received in the late 1990s from an organization that calls
itself Only Love Prevails claims the number is a mere 80,000. They
suggest that people should respond to any negative event—personally
or worldwide—by mentally chanting, "Only love prevails." When I
asked Victor Grey, author of *Web Without a Weaver* and a member of
the organization, where they came up with that number, he wrote
me: "Physicists tell us that according to the laws of wave mechanics,
the intensity of (any kind of) waves that are in phase with each other
is the square of the sum of the waves. In other words, two waves
added together are four times as intense as one wave, ten waves are
one hundred times as intense, etc. Since thought is an energy, and all
energy occurs as waves, we believe that 80,000 people all thinking the
same thing together are as powerful, in terms of creating the reality
that we all share as the 6,400,000,000 people (80,000 times 80,000)

that will inhabit the planet around the turn of the century, in their random chaotic thought. Therefore, 80,000 people all believing only in love will be enough to change the planetary reality."[2]

Could it be? Studies done by the Transcendental Meditation folks have demonstrated repeatedly that when a certain threshold of meditators is reached in a city, the city's crime rates suddenly drop. (Seven percent is the figure most often cited, although some groups claim as little as one percent.)

But even if Victor and his organization are wrong in their numbers, there is still hope. Ideas are the most powerful force in the human world: *everything* man-made originated with ideas. Our culture is an idea—the idea of domination—and it can awaken to, or *remember,* the idea of cooperation that humans lived out for millions of years.

So if you share these ideas with just one person every month, and each of them shares the ideas with one person a month, a rapid and profound multiplication of this view can spread across the world. When you do the math on this "one person a month" sharing, you discover that within less than three years every human being alive— over six billion people—could hear the message, see the vision, and feel the possibility of a better life.

Whatever the number, there is a synergistic effect in human interactions. The more people who think or believe a certain way, the more who will find it easy to think or believe that way. The more acts of mercy performed, the more people will be inclined to act mercifully. The more people turn to searching for peace and divinity, the more will be found.

New Stories Are Necessary to Change the World

 I tell you the truth, unless you change yourself and become like little children, you will never enter the kingdom of heaven. Therefore, whoever humbles himself, as this child, is the greatest in the kingdom of my heaven.

—JESUS, quoted in Matthew 18

There are those who point to methane crystals on the floor of the ocean, cold fusion, or hydrogen fuel cells as the salvation of our future. "When the oil runs low, we'll just find something else for power," they say.

And they may well be right.

But if so, these "solutions" are at best a postponing of the inevitable, and at worst could lead to catastrophe. The reason why is because they're still based in the story/myth that the goal of humanity is to dominate and conquer the Earth, that consumption is a high and positive value, and that population growth is both necessary and good.

Regardless of how much or little oil we have, it is that set of Younger Culture/dominator stories that will lead us to the wall faced by the Sumerians, Greeks, Romans, and countless others who faded into greater obscurity.

Even if there were no limit to the amount of energy we could extract from the planet, there is a limit to the size of the Earth and to the number of humans it can comfortably support . . . and we have already approached that limit, as we can see by the mass extinctions of other species and the spreading toxicity of our environment. Even if we were to move our colonization to Mars or the Moon, so long as we base our culture on the ideas of consumption, domination, and unlimited population growth, we will hit the same wall that every city-state, every Younger Culture in history has hit.

Entirely new ways of living are necessary, and if we don't adopt them voluntarily, we or our children will eventually adopt them involuntarily, and probably with great pain and difficulty in the process.

In order to make this shift, new stories are necessary, both personally and for our culture.

* * *

The culture we're born into, the role we play in that culture, our family birth order and circumstances, our race and gender, our social status or wealth: all of these things affect the stories that we tell ourselves, which define and circumscribe our experience of reality.

Those stories, because they arise from thought, are entirely personal and slightly different for every person. And they are so powerful, such strong mediators of our experience of life, that they can make us happy or sad, strong or weak, sick or well. They actually change the way our brain and nervous system work from moment to moment.

Consider the example of two people preparing to get onto a roller coaster at an amusement park:

Bill looks at the roller coaster and tells himself that it's going to be a fun ride. His internal story is that the dips and turns will be exciting, the speed and wind in his face will be invigorating, and that this is an event to enjoy. Because of this set of internal stories, when Bill is riding the roller coaster his brain will be producing endorphins and "pleasure/fun" neurochemicals. His entire body's nervous system responds to the ride in a positive and healthy way, and he steps off the ride feeling invigorated, happy, and relaxed—as if he has a "runner's high." The overall effect is that his immune system is positively activated and his body and mind are more healthy than before the ride.

Sam, on the other hand, looks at the roller-coaster cars diving down a steep hill and then swinging around a sharp curve and tells himself that it's dangerous. People have died on these things, he says, remembering the stories over the years of cars derailing or people having strokes or heart attacks from the stress. And so when Sam gets on the ride, his brain is directing his endocrine system to crank out cortisol, adrenaline, and myriad other stress/fight/flight hormones. As he steps off the car, he has depleted much of his important nutrient supply to deal with the stress, his digestive system has shut down, his blood pressure and heartbeat rate are soaring, and the strain on his body and nervous system may even have caused permanent damage.

What a difference between the stories that Bill and Sam told their respective selves about the same roller-coaster ride!

In the larger world, these cultural stories about what's real and what's unreal enfold and surround us right from birth, and are rarely questioned.

For example, Europeans lived in North America for hundreds of years before a large number of people seriously questioned the cultural story that it was a good and right thing to own slaves. Slavery is condoned in the Bible, after all, and keeping slaves is a practice that extends at least back to Gilgamesh, this ancient over-culture being the oldest and most influential of our modern cultural stories.

Because the larger "colonial American normal" culture defined Africans as being sub-human, and people like Jefferson, Washington, and Madison were born into this cultural story, few ever thought to question it. It simply was the Way Things Are, the reality of the day.

The dominant story can and does get changed, then reality changes

Two generations ago, segregation was considered both normal and realistic by most of United States culture. The story that white people told themselves back in the fifties, forties, and thirties (and before then) was that blacks were inferior to whites, and, therefore, ought to be separated from them. Many whites told themselves that this segregation by race was a kind and charitable thing for the blacks, that blacks preferred it, and that it was consistent with natural law and the teachings of the Bible.

The story of "inferior" blacks faced its first serious challenge in the United States in the 1950s when the Civil Rights movement came along, and people like Rosa Parks and Martin Luther King Jr. challenged it in such a way that it eventually became difficult for whites to continue telling it to themselves and believing it. When a large enough group of people believed the new story, our culture "shifted" and stories of black-white equality and equal opportunity became predominant.

Of course, in the face of any cultural shift there will always be keepers of the old story's flame. There are still people in Germany who believe Hitler was right. There are still whites in America who argue in favor of segregation (and even some blacks, for that matter). But because these are minority viewpoints, people who focus their lives around these particular stories are called members of subcultures or cults, and are often given labels such as racist, segregationist, or Klan member . . . whereas in 1935 they were called "good Americans" and those positions were espoused by such mainstream people as Charles Lindbergh (who publicly admired Hitler) and Henry Ford (who published an openly anti-Semitic newspaper).

The point is that our entire notion of reality, the metaphorical ground upon which we stand, our life anchors and plans, are made up of stories that can and do change over time.

When cultural stories begin to shift, the carriers of the new stories are considered oddballs, cranks, or cultists. Hitler was laughed at in the 1930s. Southern governors in the 1960s turned fire hoses and police dogs on blacks demonstrating for civil rights. The Romans fed the early Christians to lions. Washington and Jefferson were referred to in the British press and among large parts of the American population prior to the Revolutionary War as misfits and malcontents.

But stories change when a critical mass is achieved. Carl Jung speculated that this was connected with a process involving the collective unconscious; Rupert Sheldrake calls it morphic resonance. Richard Brodie, in his book *Virus of the Mind*, calls these new cultural stories *memes* and points out that if they're sufficiently "contagious," the new memes will eventually infect an entire culture and become part of the shared view of reality of that culture. As a result, the culture itself changes along with the people within it.[1]

So let's examine some more new-but-ancient stories. . . .

Touching the Sacred

 Then I saw that the wall had never been there, that the "unheard of" is here and this, not something and somewhere else, that the "offering" is here and now, always and everywhere—"surrendered" to be what, in me, God gives of Himself to Himself.

—DAG HAMMARSKJÖLD (1905–1961),
secretary general of the United Nations
1953–1961, diary entry, 1954

I remember a summer when I was five years old. My parents had recently purchased a hammock and put it in our backyard, and I was lying on it on a bright sunny afternoon. The sky was a deep blue, with thin wispy clouds, and I could smell the fresh-mowed grass crushed by the green-painted metal frame of the hammock. I could feel the ropes of the hammock against my back through my T-shirt and pressing against my bare legs below my shorts, and hear the melodic sounds of birds singing in the trees that surrounded the yard. One of the birds was repeating over and over a three-note call, while others chirped randomly.

I stared at the sky, noticing little specks in my field of vision and how they'd jump when I moved my eyes and then slowly settle when I held my sight on a particular bit of cloud. There was a gentle wind blowing, and I could hear it rustling the leaves of the huge old maple tree about thirty feet from me; the hammock rocked very slightly, a soothing motion that made the sky seem to tilt slightly from side to side.

I took a deep breath and noticed how breathing deeply seemed to brighten the sky, and smelled the blooming roses and hollyhocks and flowers along the edge of the yard, mingled with the fresh-laundry smell of the pillowcase on the pillow under my head. My fingers interlaced across my stomach, I felt the warmth of the sun on my bare arms, legs, and face.

Turning my head to my left, I noticed that I was 10 feet from a stand of pink, white, and yellow hollyhocks, covered with blossoms and standing 5 feet tall. The thick white stamens erupted from the waxy, colored petals, and honeybees and bumblebees moved lazily from flower to flower gathering pollen. I could hear them buzzing, as if they were humming their pleasure at finding the pollen.

As I looked at the way colors flowed from pink to white on the flower petals, noticed how the sounds of the birds had changed with the movement of my head, felt the sun now full-warm on the right side of my face, I was washed over with a sense of total Now. I saw that the flowers were alive, the bees were alive, the tree and the birds were alive, and I was alive. The air was crystal clear, and I noticed the empty space between me and the flowers, the distance between me and the grass, the next house over, and the tree. Even the empty spaces vibrated with life.

"Wow," I said softly, then heard the sound of my own voice, and that was another miracle, amazing me all over again. It was a perfectly ordinary moment, yet filled with Spirit.

In its simplest and in its most complex forms, that is one of the most powerful forms of meditation, a touching of the presence of Life itself.

Einstein wrote about how past and present are only concepts we form in our minds but having no ultimate reality. Everything that exists and happens is only in the ever-constant *now*, and *now* is the only time that exists.

Einstein also said that he rarely thought things through with his intellect, but instead achieved his most important realizations in flashes of insight, moments of intuitive *knowing*. He was describing, both in his concept of time and his descriptions of how he came to see new concepts, a form of meditation.

Viewing the past

If you look back over your life, back over the years you've been alive on this planet, year by year and decade by decade, you'll probably notice there's a vast gray sea of recollection, and embedded in it are a few crystal clear moments of vivid memories.

These remembered times often seem idiosyncratic: Why would I carry around all my life the memory of an afternoon on a hammock? Or walking down the street in New York City in 1973, or sitting by a pond when I was 16, or Miss Hemmer's biology class in the seventh grade?

What's so special about them? Why are those memories so much more vivid than the "really important" things, the things I *wanted* to remember, such as how to do quadratic equations, the name of that reporter I'm meeting in an hour, or the directions to a place where I have to give a speech?

Sometimes, the things we choose to remember make perfect sense: who could forget their wedding, or the birth of their children, or their first day of school?

But the idiosyncratic and the reasonable memories share something in common, and that something is at the core of the meditative state: it's what I call *presence*.

If you re-examine any of those memories from your past—big or little—the one thing you'll find they all share is that at the moment a particular memory was imprinted in your brain, you were not talking to yourself in your head. You were not thinking or worrying or imagining or comparing or judging: you were *being*. You had set aside the stories, and were only experiencing.

Lying in that hammock, as I felt the sun on my skin, heard the birds and the breeze, and saw the life in those flowers, I was so shocked by the vitality and reality and aliveness of it all that I stopped thinking about it for a moment and simply experienced it. I was there, then. This sense of presence is at the core of the meditative and the mystical experience. It is the time when we are not thinking, but are instead alive and aware.

Achieve presence

Different people achieve this by different routes, but all methods have the effect of shutting down the thinking apparatus, which then allows our true consciousness to wake up and look around and see, hear, feel, taste, and smell the world.

Saint John of the Cross, for example, had a particularly difficult route to this experience. Born in 1542 in Fontiveros, a town in the Castilian region of Spain, he was the son of an impoverished weaver who was forced to convert from Judaism to Roman Catholicism. His father died when he was very young, and Juan helped support his mother by begging and working with her on her loom. Around the age of 21, he joined the Catholic order of the Carmelites and took on the name of Juan de Santo Matías.

Shortly after this time, he met Teresa of Avila, another Spanish mystic, who was trying to reform the Carmelite Order, moving them toward vows of poverty and mercy and away from the pomp, glamour, and power of the church. She was in her fifties at the time, and enlisted this young man's aid in her reformation of their order.

Because of Juan's support of Teresa's reformation, he was arrested by the church and kept for a year in a cell made from a converted cupboard. There was no light most of the day, and he couldn't stand up straight. He was not allowed to wash or change his clothes for six months, although he was infested with lice and fleas, and every day during this period of his imprisonment he was subjected to the church's "circular discipline." Daily, he would be removed from his cupboard and stripped of his shirt. Scraps of bread, a cup of water, and an occasional sardine were thrown on the floor, and as he kneeled to eat them, a group of monks walked in a circle around him, bursting the skin on his back with leather and wooden whips. They stripped the skin from his back and shoulders so many times, occasionally also breaking his shoulder and rib bones, that he was crippled for the rest of his life.

After six months of this, his lashings were reduced to once a week, lest he die from loss of blood. A new jailer took mercy on him and gave him paper and pen, and allowed the door to the cupboard to be open far enough to let in some light from the room so he could write.

It was during this time that he wrote some of his most profound and insightful works, including *Cántico Espiritual, La Noche Oscura*.

Consider this stanza from his poem *"Sin arrimo y con arrimo"* ("Without Help and With Help"), about how he was touched ("helped") by divinity even in his moments of greatest darkness:

> *Without help and with help*
> *Without light and living in the darkness*
> *Everything consumes me.*
> *My soul is in threads.*
> *From everything, something is grown*
> *And uplifted by itself*
> *Into a life filled with ecstasy and richness.*
> *Only a being God helped*
> *For that reason, it will be said,*
> *The thing I most cherish*
> *That my soul see itself even now,*
> *Without help and with help.*[1]

John used his privations and pain as a tool to turn off his thinking mind. In that quiet place—which he wrote about extensively in *Dark Night of the Soul*—he met the love, light, and presence of God. It was his form of meditation.

When we understand that this—finding that quiet place within where *thinking* ends and *consciousness* begins—is the most important goal and purpose of meditation, then it's easier to understand and use the various forms of meditation.

Nearly every spiritual tradition on Earth has developed some form of meditative practice, and each is intended to arrive at the same place. Because each practice is rooted in the culture and assumptions and traditions of a particular time and place in the world, each has a different flavor and energy.

While many books and teachers will tell you that meditation is about reducing your blood pressure or calming your jangled nerves or improving your health, those are all just side effects. They do happen, as has been confirmed in study after study, and meditation can be a powerful tool for physical or emotional healing . . . but that's not where its real value rests. The true power of meditation—and the reason for meditating—is to become awake *in this very moment*. And from that place—that here-and-now touching of the power of life— we can find the ability to transform ourselves and others in ways that can and will transform the world.

This seemingly very personal work is actually among the most important things we can do to save the world, because as we become grounded in the present, we gain the power to create change. We also acquire and radiate a spiritual strength—the solidity and reality of spirit that tribal people have known about and used for millennia.

It's amazing to think that it's possible to change the world by changing ourselves, by changing the way we think, live, and experience every moment, but that's been the core message of almost every religion in history, from the most ancient and primal to the most modern. You can change and save the world by changing yourself. And that begins with waking up to the power of life in the present, and finding there the presence of your Creator and all creation.

Learn to Create Awareness

 The millions are awake enough for physical labor. But only one in a million is awake enough for effective intellectual exertions, only one in a hundred million to a poetic or divine life.

To be awake is to be alive.

I have never yet met a man who was quite awake. How could I have looked him in the face?

—HENRY DAVID THOREAU, *Walden*, 1854

Most of us walk through daily life, driving down the street, sitting in the office, or wandering around the house, in a state of disconnection from the natural world around us. We're thinking of the past or the future, worrying about some problem or task, preparing to meet or avoid the challenges of our day. In a sense, we're not alive: the experience of aliveness is not a normal part of our daily lives.

For many people in modern cultures, aliveness comes most vividly when they're outdoors experiencing nature. The more unusual and extraordinary the natural sight (waterfalls, mountains, canyons, redwood forests, oceans), the stronger the feeling many people get of spiritual reality.

But this is a rare and only occasional experience. Wouldn't it be wonderful if we could experience that sense of wonder, awe, and connectedness anytime we wanted to?

The Kogi Indians have developed a technique to do this. While it's not what I'm suggesting here (I'll share with you a less drastic method in just a few pages), it shows us the cultural importance of having members of a society who view their spiritual work as real and visceral, rather than thinking of themselves as interpreters of a revealed law or wielders of power over a congregation. It's the story of how a tribe has learned to have at least one among them who sees divinity all around him, at all times.

The priests of the Kogi learn by divination when a "High Soul"—a person destined to be a priest—is about to be born into their tribe. This child is taken shortly after birth into a deep cave, where his mother goes every few hours to feed him and attend to his needs. He can see only enough light to keep his eyes developing, and can hear only the sounds of the cave to keep his ears alive. As he grows, the nurturing and feeding tasks are taken over by a group the Kogi priests called the Mamas (no relation to the word "mother"), who begin describing to the young boy what he'll see, hear, and feel when he finally steps out of the cave for the first time.

They tell him stories of the Great Mother, who they believe is the creator of the Nine Worlds, and how this particular world in which we live is large, beautiful, and rich in detail. The boy, seeing only dim light inside the cave and occasional candlelight, can only imagine what it must be like "out there." He wonders what a tree or mountain must look like, how moss could grow on rocks, how these mystical animals can fly through the air, and what it must feel like to have the Great Mother's sun warm his body.

When he approaches puberty, he is brought out of the cave with great ritual and allowed to see the world for the very first time.

What a shock! How astounding! Look at the detail in that leaf— how did the Great Mother know to do that? Look at the distant mountains—how is it possible for something to be so large? And the trees and the flowers and the animals and the birds—he looks around, listens to the world, feels and smells the fresh air, and the experience of awe and splendor and gratitude is so great that the boy most often falls to his knees before the majesty of the Great Mother's creation.

For the rest of his life, every time he opens his eyes he will see anew the handiwork of the Creator of the world, and he will never cease to be washed over with joy and awe and respect for Her handiwork.

Because of this unique perspective he has of the world, he plays a vital role in the tribe. He's the tribe's memory of divinity, its connection to the Spirit World, its conscience when anybody would think to stray into activities that may harm the Great Mother's beautiful and detailed work.[1]

For many modern people, the closest we come to this is looking up into the vastness of the night sky, a sight that is not even part of most people's daily experience.

Do you remember a time, as a child, when you stood outside on a warm or cold evening, alone and in awe, looking up into the depths of the universe and seeing all those thousands of twinkling points of light? Each one is a sun, just like ours, and may have planets revolving around them. Their distances from us are unimaginable, and far beyond the faintest, most distant star we can see, there are a billion billion more.

Remember that time for a moment, seeing what you saw, feeling what you felt, and hearing what you heard. Now, as you finish this sentence, look up from this book and glance around you at the place where you are now.

Here's a secret that is mostly known only to physicists and astronomers: you were just then looking at the remnants of a dead star (as you are now). When our universe first came into being, all of space was filled with subatomic particles. Through the force of gravity and over millions of years, these congealed into atoms of hydrogen, filling space with that gas. At that time, hydrogen was the only element that existed. Gravity pulled together clouds of hydrogen, and caused them to collapse into themselves and become very hot from the pressure. This heat "ignited" these clouds with the fire of nuclear fusion, and stars were formed.

Deep within the heart of one of those early stars, nuclear fusion reactions caused two hydrogen atoms to combine into a helium atom, producing the second element in existence and releasing huge amounts of heat and light from the atomic reaction. The pressures were so great that the process didn't stop there. Atoms of helium burned/fused into larger and heavier atoms, and every element we know of in the periodic table eventually came to fill the core of the star—iron, carbon, gold, boron, oxygen, neon, argon, nitrogen, calcium, potassium, and the dozens of others.

As this process continued, the core became heavy with these new elements and began to cool, causing the color of the star to shift toward red, and the outer edges of the star to expand. This was the first step in the death of that far-distant star, the one whose fiery core-remnant you are holding in your hands as this book.

Over the next few hundred million years, the star continued to balloon until a critical point was reached where it no longer could hold itself together, and it exploded, spewing the matter created in its core out over billions of miles of space and destroying almost everything within its immediate vicinity. This process is called a supernova, and signals the death of a star.

This is where *all* the matter of our world (except hydrogen) came from: it was created in the heart of a star. Not only that, the star had to die for those elements to reach us.

After the star exploded, some of the matter it threw out into space was drawn together by the force of gravity and formed huge clods of physical material. The larger of these became what we call planets, and they began the slow process of cooling down (our Earth took about five billion years). Flying through space, most were captured by the gravitational pull of another, still-living and burning star, and began to orbit their new host. From their new sun, they derived warmth, which in the case of Earth drove the process of photosynthesis and led to the life we see around us, as the elements created in the death of a far-distant star were drawn up into plant and animal matter.

Remember your awe at looking up into the night sky so long ago? Remember how distant those stars seemed, and how incredible you thought it would be to be able to see one up close or touch its matter? Now look around your room again and realize, as you do, that everything you see is matter that was created in the heart of a star, was blown out into black, empty space when that star died, and has come to be accumulated here as what we call the planet Earth and everything on and in it.

Not only is the matter around you star-stuff, but you are, too. There is not a single cell in your body that is not made of matter formed in the heart, and then the death, of a distant and now-extinct star. Like the young Kogi priest who sees the hand of divinity every time he glances at the world, you, too, can see the hand of divinity— of a power and force greater than any of us can imagine—simply by looking around, by touching the pages of this book (which are also made of star-stuff), by hearing the sounds in your room (which are star-stuff vibrating and the sound vibrations are being conducted through star-stuff air to your star-stuff eardrum).

If you do this exercise a few times over the rest of this day or evening, and then again tomorrow, you'll discover that your point of view, your perspective, will change. This is another step in opening your eyes. More follow.

Lessons from a Monk

You used the word "civilization," which means a set of abstractions, symbols, conventions. Experience tends to be vicarious; emotions are predigested and electrical; ideas become more real than things.

—Jack Vance (b. 1916), *The Gray Prince*, 1974

The best way to remake the world is by starting with yourself and your own internal world.

I was reminded of this in a completely unexpected encounter, in an unexpected place, with a rare soul who became an honored friend.

In 1985, I was a featured speaker at a three-day conference sponsored by American Express and KLM Airlines for people in the travel industry. It was held at a beautiful hotel in downtown Amsterdam, and after one of my presentations on communication skills, a few hundred of the attendees and some of the presenters were invited to a gala dinner put on by the Dutch Tourism Board.

As I walked into the Grand Ballroom, filled with round tables and seating two or three hundred people, I spotted an Oriental-looking man at a distant table with an empty seat beside him. He waved me over, jumped to his feet, and introduced himself as Dr. George Than. He stood about a foot shorter than I, and was thirty years older, although I would have guessed he was in his forties rather than his sixties. George told me he was originally from Burma. He had the broad face and dark skin of most Burmese, and spoke and gestured in an elegant and purposeful fashion, like a member of royalty or one who was trained in diplomacy. His black hair had a few speckles of gray and his warm smile seemed to move every muscle in his face.

I liked him as soon as I saw him, but I felt there was more to it than that. It was an odd feeling, actually, one I don't often experience when meeting people. I felt that he was *grounded*. There was a presence to him, a solidity, an "I am here and it is now" that, in my memory, makes the image of him bright and vivid and the rest of the room dim and out of focus.

In addition, I was certain that I knew him from somewhere. I felt so sure of it that I said, "Where have we met before, George?"

He laughed, looked right into my eyes, and said in a matter-of-fact tone, "Maybe we were monks together, a long, long time ago."

We spent much of the rest of the conference together, talking and walking around Amsterdam. George's wife, Nancy, was a travel agent, which is what had brought him to the conference, and he practiced urology in Salinas, California. Just after World War II, he'd received his M.D. degree and medical license in the United States. Shortly afterward, he'd been drafted by his brother-in-law, who until recently was the president of Burma, into the Burmese foreign service.

"One day I woke up in London," George told me, "and realized that I really didn't want to be a diplomat. I wanted to practice medicine. So I went back to my brother-in-law and told him that I wanted out. But he wouldn't let me out: he said that if I left the country, he'd declare me a persona non grata and banish me from Burma because I knew too many state secrets. So I went to Thailand and entered a monastery."

George entered the Wat Sri Chong monastery in Lampong, Thailand, and began to study Sattipattana ("attentive observation" or "mindfulness") under a teacher named Uwaing, named after the Bodaw Uwaing, a legendary figure in Thai Buddhism.

Meeting George was an important reminder to me of the need to be awake and aware in the moment. He practiced daily the meditation he learned in the monastery, as well as t'ai chi ch'uan, a moving form of meditation.

A few years later, two weeks before George's sixty-fifth birthday,

I called and asked him what his plans were for his birthday. "I have none," he said, "but I would like to go someplace and meditate. Can you join me?"

I called my travel agent and booked a cheap flight to San Francisco, and on George's birthday together we drove up north of San Francisco to Mount Tamalpais, through redwood and eucalyptus trees, to a point that overlooks the Pacific Ocean. From there, we sat and practiced his Sattipattana and Vipassana meditations as the sun turned crimson, then fire-red, and sank into the ocean. It was a wonderful day.

In January of 1997, I received a phone call from George; it was the first time we'd talked in two or three years. "I have cancer of the gallbladder and liver," he said, "and it's inoperable. But I'm seventy-four years old, so maybe it'll move slowly."

The news stunned me, and I immediately booked a flight to San Francisco.

George was having a "good day," according to Nancy. He could get out of bed and walk around. In fact, he was nearly jumping, he seemed so excited that I'd come to Salinas to see him. "It's so good that you came!" he said several times.

Given the gravity of his situation, his adult children and relatives were coming in to visit from as far away as London, so we went out to a Japanese restaurant for lunch. Drinking green tea and slurping miso soup, I asked George how he was spending most of his time.

"I'm spending more and more time in the void," he said, smiling.

"The void?"

"Normally when I meditate, I bring my attention to the present moment, to mindful awareness. Do you know this place?"

"Here and now," I said.

"Exactly," he said. "The moment you have a thought"—he snapped his fingers—"poof, it's gone and you've popped out of the present and into the thought. So usually I practice Sattipattana. Just being present, but not judging things, not thinking about them, just being here."

"How did you first learn that?" I asked.

"In Wat Sri Chong, I had a small cave where I'd sit every afternoon. My teacher told me to simply be there, to just sit there, and to count my breaths. He said I should never count higher than four, because most people cannot sustain their attention for the time span of four breaths. So I count four breaths, and then begin over at one."

I told him how when I'd first learned Zazen, or Zen sitting meditation, I'd been taught by the roshi to count ten breaths. That brought a huge laugh from George.

"If you can stay with your breath for a full ten breaths, you are very close to enlightenment! I would never suggest that long for a beginner."

"So what happened when you sat and just counted your breaths?"

"Well, at first, of course, thoughts would bubble up. Over and over again. Think about this or that, the cave and the monastery, my country and its government, what was for lunch: all sorts of things. Then I'd think about my breathing and my body." He smiled. "That's a difficult one to avoid."

"And then?"

"Finally, after just a week or so of this, one day I went to sit after lunch, and when the gong went off at six P.M. for dinner, I thought it had only been a few minutes. The time went so fast, because I was just there, then."

"Do you think that when people are meditating like this, time moves at a different speed?" I said.

George shrugged. "I don't know. It does for me. Maybe time isn't even real; maybe it's just an artifice of our brain, something that we create to divide this from that. But when you sit and are present fully in the now, there is no time. Only now."

"So you did that for years?" I said.

He waved a chopstick at me. "I still do it, many times daily. But more and more, after that time decades ago, I learned to bring that sense of nowness into my daily life, into my work as a doctor, into everything I do. That is the true meaning of Sattipattana, to live mindfully. The sitting part is just meditation, but the really powerful part is Sattipattana, or living with attentive observation. Whether you're eating, walking down the street, performing surgery, or talking to your family members. Always observe yourself. And then observe yourself observing yourself. Eventually you get close to that place."

"And what about that void you said you've been spending a lot of time in?" I asked.

"Often when I meditate with my eyes closed," he said, "I will slip into the void. This is not a no-place. It's the all-place, the place from which the universe came into being. And when I slip into it, I become one with the universe. It's an extraordinary experience, and I'm finding myself drawn to it more and more as I approach the end of my life."

"How do you do it? How would a beginner approach that?"

George shrugged and put a piece of tofu into his mouth, chewed it thoughtfully for a moment as if it were the most important act he'd ever performed, then replied, "I'd suggest that they use the tool that suits them best. If they're tactile, I'd say they should use a rosary or mala beads. If they're auditory, then they should use a mantra, such as 'Om,' or 'Amen,' or the name of a holy person they feel particularly close to. If they're a predominantly visual person, then I'd suggest that they hold fast to a vision of a saint, or Christ if they're a Christian, or a teacher of theirs who's no longer on this Earth."

"How about using a candle?" I asked, thinking of how my old mentor Hamid Bey had been trained in the Coptic Order temple in Cairo.

He nodded. "That would work for a visual person. Or a mandala. Most important is that they should choose something that works for them, that they like, that they're comfortable with. These people who try to say that there is only one way to do this, that everybody must do it their particular way, are not allowing for the differences between humans. There are many paths, many methods."

"And what do you do with the sound, the beads, or the sight?" I asked.

"You hang out with it," he said. "Some teachers say you must sit up straight, the spine must be straight, all that. But I can't do that now because it hurts my liver so much when I get into some of those postures, and an old man like me may have arthritis or some other problem with getting into a sitting posture." He moved his body toward one of the postures, then winced. "These things are not so important. I lay down on my bed now, on my back, and work with the rosary my teacher gave me almost fifty years ago before he died. It takes me to a place of single-mindedness, of concentration, of focus."

"Is that different from attentive observation, Sattipattana? It seems like that's the opposite, a kind of non-focusing."

"No," George said. "You must first learn to focus, to concentrate. You can learn that, and you do that with the rosary or the sound or the vision. And then, when you're practicing Sattipattana, you bring that power of concentration to focus on the present moment. So first you must learn to concentrate."

"And the void?"

"Well," he said with a faint smile, "it used to be that when I'd concentrate I'd just become totally absorbed in the feeling of the beads, the sound, and the vision. But now I find that very often I'm stepping past them, so to speak. Going beyond them—beyond even the beyond. And when I step through into that place, I find myself in the void. I don't know how to describe it beyond that, other than that it is very wonderful and very powerful. I bring it back into my daily life, then, and it strengthens me."

The best way to remake the world is by starting with yourself and your *internal* world. I urge you to experiment with the meditations that George describes. For example, most people believe they "experience the world." We see what we see, hear what we hear, feel what we feel, taste and smell what is there to be tasted and smelled. But in actual fact, very few people can or do experience true sensory reality.

Instead, what we do is take a sample of sensory input, and then immediately begin to conceptualize it, discussing it with ourselves in our own heads. This internal dialogue, made up mostly of judging, evaluating, comparing, and associating, pulls us a step back from the sensory reality that surrounds us. Yet, we *think* it's reality. We're not even aware of the extra step everything goes through before it gets into our thoughts.

If this seems abstract or didactic, try this simple experiment, a reminder of what you were invited to do a few chapters back, to hear the noise in your head. After reading this paragraph, put down this book for a moment and look around you. Listen to what you can hear, then feel the sensations on your skin and in your body, and then see what you see. And try not to discuss it with yourself in your head. If you can maintain "no words in the head" for even three seconds, totally "here" with the sensory world around you, you're doing better than most beginning meditation students. If you can hold that state of pure sensory awareness for more than a minute or two, you've achieved the state of mind that the contemporary Christian monk the late Thomas Merton often taught and wrote about.

So what happens is that we're constantly flickering between experiencing the world through our senses and then discussing with ourselves that sensory information. That discussion pulls us back from the real world, from the experience of life, and drops us into our own individual consciousness of thinking, judging, and comparing.

This internal thinking headspace of consciousness is where we spend most of our lives. The way we react to the world around us—the type of internal conversation we have with ourselves—is based on the stories we tell ourselves about "reality." By turning off the chatter, the evaluations and judgments, we can let the stories drop away and experience a moment of reality, a moment of the sacred now.

Find the path—auditory, tactile, visual—that works for you, and learn to experience "now-ness." Sit quietly and simply listen, see, and feel—letting the internal chatter dissipate—for 10 or 20 minutes each day. Since you and I and every other living thing are interconnected, your meditation will positively affect yourself and your reality, and send ripples of change throughout the world.

Re-empower Women

We have now traced the history of women from Paradise to the nineteenth century and have heard nothing through the long roll of the ages but the clank of their fetters.

—LADY JANE WILDE (1821–1896)

A friend who is a psychiatrist with training in neurochemistry once joked to me, "The most dangerous drug in the world is testosterone." History suggests he's right.

Exhaustive analysis of "pre-historic" cultures, such as done by Riane Eisler and others, indicates that in almost all Older Cultures women had equal status with men, and in a few they were even in charge. One theory for this is that women uniquely bring life into the world, and it may not have been until humans moved from hunting/gathering to herding/agriculture that they began to understand genetics. Women ran the show because they controlled life itself, producing life from their bodies.

When everybody figured out that the men had a role to play in the process, however, during the early herding times, some of the men pulled off a power-grab, converting the gods that were worshipped from female to male, and asserting control over the fertility of women the same way they controlled the fertility of a field or a flock of sheep. The men took over and testosterone-driven behaviors came to dominate the beginnings of our Younger Culture: aggression, competition, domination, warfare.

In the early 1990s when European missionaries taught Australian Aborigine hunter-gatherers how to play "football," the children played until both sides had equal scores: that was when the game was over, in their mind. The missionaries worked for more than a year to convince the children that there should be winners and losers. The children lived in a matrilineal society that valued cooperation; the Englishmen came from a patriarchal society that valued domination.

Only women in the Iroquois tribe could vote on most issues. As a result, decisions regarding relations with other tribes were more often made in the context of "What will work for our children?" rather than "Who wins?" or considerations of pride, power, or conquest.

Similarly, we find that populations are exploding in almost every nation of the world where women are dominated, treated like cattle or goods, or exploited and controlled. The men in these countries are making the decisions, and one of the male values is "Have many sons to build the biggest army" (another common one is "Have sex whenever you want, with whomever you want").

On the other hand, in those nations where women have relatively equal position and power with men, there are lower birth rates, often even to the point of zero population growth, as has been achieved in many of the countries of northern Europe. In almost every country we can see this equation demonstrated: male domination equals population explosion; relative male-female equality equals sustainable population.

In this regard, you could say that the women's rights movement is truly a *human* rights movement. So another solution to this mess we find ourselves in is to give power back to women in all realms, including the social, familial, religious, military, and business worlds.

The Secret of "Enough"

 For I have learned, in whatsoever state I am, therewith to be content.

—ST. PAUL, Letter to the Philippians, 4:11

First, the truth.

If you are naked, cold, and hungry, and somehow you get shelter, clothing, and food, you will feel better. Providing for these necessities creates a qualitative change in life, and could even be said to, in some ways, produce "happiness." You feel comfortable and safe. Your state of mind and emotional sense of well-being have improved as a result of these external changes, the result of your having acquired some stuff. Let's refer to this as the "enough point." It represents the point where a person has security, where their life and existence is not in danger.

Now, the lie or myth.

"If some stuff will make you happy, then twice as much stuff will make you twice as happy, ten times as much will make you ten times as happy, and so on into infinity."

By this logic, the fabulously rich such as Prince Charles, Bill Gates, or King Fahd must live in a state of perpetual bliss. "Greed is good," the oft-repeated mantra of the Reagan era, embodied the religious or moral way of expressing this myth. More is better. He who dies with the most toys wins.

Many Americans who lived through the Great Depression discovered in that time that "more is better" is a myth. My wife's grandmother, who died in her nineties and was still living frugally but comfortably, owned a farm during that time, and was able to provide for nearly all her family's needs by growing her own food, burning wood, and making their clothing. Recycling wasn't a fad to save the environment, but a necessary part of staying alive and comfortable. Great-grandma had enough money in investments and from the sale of the farm to live a rather extravagant lifestyle, but she still bought her two dresses each year from the Sears catalog, collected rainwater to wash her beautiful long hair, wrote poetry, and found joy in preparing her own meals from scratch. She saw the myth for what it was, and was unaffected by it.

Some, of course, came through the Great Depression so scarred by the experience that they went in the opposite direction and totally embraced the myth. The excesses of Howard Hughes, for example, are legendary—as is the painful reality that almost limitless resources never bought him happiness.

Similarly, the myth has become a core belief in the cultures of America, much of Europe, and most of the developing world. Advertisers encourage children and adults to acquire products they don't need, with the implicit message that getting, having, and using things will produce happiness. Often the advertising message of "Buy this and you will be happier" is so blatant as to be startling to a person sensitized to the myth. Forget about the "enough point," these sellers say: this product or service will be the one that finally brings you fulfillment.

The meaning of wealth

But we are the people—both those who feel that "enough" is a humble level of comfort and those who crave great wealth—of our culture. Like the air we breathe, it's often easy to forget that we are members of a culture that is unique and has its own assumptions. This Younger Culture is based on a simple economy—you produce goods or services that have value to others, and then exchange them with others for goods or services they produced that you want or need. Money arose as a way of simplifying the exchange, but this is the basic equation. This concept of wealth as a measure of goods or money owned is intrinsic to these cultures, and so in this regard you could say that all of these different cultures around the world are really the same, variations on a single theme, different patterns woven of the same cloth.

The wealth of security

While they number fewer than 1 percent of all humans on the planet, the result of a relentless five-millennium genocide by our worldwide Younger Culture, members of Older Cultures are still alive on Earth. There are also people whose Older Culture ways have been so recently taken from them—such as many Native American tribes— that while they may no longer live the Older Culture way, they remember it.

In these Older Cultures, the concept of "More is better" is unknown. They would consider "Greed is good" to be the statement of an insane person. One person eating near another who is hungry is an obscene act.

These values and norms of behavior are quite different from those we see in our own world today. But why? The reason is simple: security is their wealth, not goods or services.

In Older Cultures, the goal of the entire community is to bring every person in the community to the "enough point." Once that is reached and ensured, people are free to pursue their own personal interests and bliss. The shaman explores trance states, the potter makes more elegant pots, the storyteller spins new yarns, and parents play with and teach their children how to live successfully.

But aren't they dirt-poor?

Because Older Culture people usually work together to create enough food, shelter, clothing, and comfort to reach the "enough point," and then shift their attention and values at that point to other, more internal pursuits (such as fun or spirituality), they appear to us in Younger Cultures to be poor.

I remember spending a few days with a Native American healer who taught me about a particular ritual I've promised not to reveal in my writing. He lived in a mobile home in the desert, on reservation land that was pitifully lacking in anything except scrub brush, cactus, and dust. His car, a 1970s Chevy, was missing major body parts, and he traded healing ceremonies with the locals for food, gasoline, clothes, and nearly everything else he needed. His income in actual cash was probably less than $500 a year, and if you added up the total market value of everything else he took in during the course of a year, it was probably less than $5,000. By any standard of contemporary Western culture, he was about as poor as you can be in America and still stay alive. And his lifestyle was nearly identical to the other two hundred or three hundred families who lived within twenty miles of him and were members of his tribe: they were all "poor."

But he had things that most of the people I knew back in Atlanta living in upscale suburban homes lacked.

If he became ill, people would care for him. If he needed food or clothes, they'd give them to him. If he was in trouble, they were there with him. When his only child needed something, somehow it always materialized from the local community. When he got old, he knew somebody would take him in; if he lost his home, others would help him build or find another. No matter what happened to him, it was as if it happened to the entire community.

As we got to know each other, and I met people in his small "town," I discovered that his riches of security and support from his neighbors weren't unique to him because he was the community's healer. The same was true of every person in the "town," from the guys who did part-time carpentry work in the Anglo city 122 miles away to the town drunk(s). *Everybody* had cradle-to-grave security, to the maximum extent that it could be provided by the members of the tribe.

Our poverty

After returning from my first trip to New Mexico, I had dinner with a friend who is a successful attorney with a big law firm in Atlanta.

"What would happen if you lost your job?" I asked him.

He shrugged. "I'd probably get another one."

"What if the job market was bad? If there was a recession or depression? Or what if you lost your job because of some monumental screw-up you did on a case?"

He looked at his plate of spaghetti with a troubled expression, staring at the twisted strands as if his future were there. "I don't know," he said softly. "I suppose I'd lose my house first: the mortgage payments, insurance, and taxes are well over two thousand dollars a month. And the car is another five hundred."

"And if your health went bad?" I said. "If you had some serious disease?"

He looked up. "You mean without the insurance from my employer?"

"Yeah."

"I'd die," he said. "I have a colleague who spends most of his time defending insurance companies who've done that to people who got sick. They then start looking through the insurance applications to see if there was anything on there the people forgot to mention when they filled it out, like a pre-existing condition, or that they'd once been turned down for insurance. If they find it, they dump them. I know of several people who've died, who could be alive today if they'd had the money for the medical care."

"And when you get old?"

"I have my retirement fund. My 401(k)."

"What if your company ripped it off, or it was all in stocks and the market crashed?"

He shook his head. "I'd be living on the streets, or in my kid's garage, assuming he could afford to have me. It wouldn't be pretty."

Even more than his words, his tone of voice and his eyes gave away his essential insecurity. If his employer went down, so would he. He was living—as was I at the time—on a tenuous thread of debt, workaday income, and hope that the government could somehow manage to keep the country's financial house of cards from crashing down as it had so many times in the past few centuries.

"If you could have anything at all," I asked him, "what would it be?"

"That's easy," he said, smiling. "More time. There aren't enough hours in the day, and I feel as though I'm on a continuous treadmill. There's never enough time to spend with my kids, my wife, our family and friends, or even to read a good book. Three nights out of the

week I bring work home, and I know that if I'm going to make part-
ner in the firm one day, it'll have to become five, and maybe even
seven days a week. I have no time."

My friend, surrounded by a wealth of physical possessions, a
fancy home with elegant carpeting and furniture, a new Mercedes,
wearing an $1,800 suit, was steeped in the poverty that is unique to
Younger Cultures: the poverty of spirit, of time, of security and
support. His life had no safe foundation, and seemed to have little
meaning beyond achieving the next level of income and creature
comforts.

As my Native American mentor said of me, "Boy, you think
you're rich, but you're poor beyond your imaginings." So we must, as
a culture, rediscover where the point of "enough" is, both materially
and spiritually. By finding this point, you become infinitely richer.

Respect Other Cultures and Communities

 But lo, thou requirest truth in the inward parts: and shall make me to understand wisdom secretly.

—PSALM 51:6

In the context of the many "rainbow" movements in the United States and elsewhere that speak of embracing all people as "part of our family," what follows may seem heretical, but that's certainly not my intention.

In tribal life, differences between cultural and racial groups are recognized as entirely sensible and natural. Many voices, from the leaders of the Black Muslims to numerous Native American tribes, have been raised to say, "We don't want to become part of your culture—we want our own, unique culture." The most common response to such voices has been to label them as racist, culturist, or elitist.

Is it any less racist, culturist, or elitist when one cultural group says to another, "Abandon your old ways, your ancient or modern traditions, your unique language and religions, and join our society"? In many ways, the call for "inclusiveness" in modern society is really a friendly mask put over the twisted face of the dominator city-state Younger Culture that has worked so hard for so many years around the world to destroy indigenous peoples and their traditions. It's a form of coercive *cultural* evangelism.

Many people are completely unconscious of this, because it's part of today's dominant story—it's the way things have always been, as far as they know. But notice: it carries implicitly the message "We'll gladly make room for you to join our culture, because ours is so much better than yours."

While cultural diversity is anathema to the traditional city-state, it's essential to the survival of tribalism.

The loss of diversity in any system increases the vulnerability of that system. Single-species forests fall easy prey to beetles or fungi; single-hybrid crops are overly sensitive to changes in temperature or moisture; single-source power grids are in danger of disruption and collapse if a power plant breaks down or an accident happens.

Similarly, with human societies, we must share the message that tribal peoples play an important role in the human ecosystem. They're not just here to show "civilized" city-state dwellers which jungle plants may cure cancer: their continued existence is necessary to maintain the cultural and genetic diversity of the human species.

Tribal people have almost no defense against the predations of the city-state and its poachers. They are being exterminated across the planet, and even those who are no longer the subject of bounties or organized hunts are now being herded onto reservations, given Westernized food, clothing, and "job opportunities." Many are losing their original knowledge, skills, and spirituality to consumerism, television, and organized religions.

"Live and let live" is a statement of tribal values, not city-state values. Different tribes will have different values and lifestyles, some quite different from ours. They will look, behave, and even talk differently. They will practice different religions, eat different foods, create different types of places to live and sleep, and wear different clothing. This is a good thing, from their point of view.

Then again, the ongoing regulation of the Native American religions by the U.S. government is a good thing from the point of view of the city-state.[1] Native Americans who share "American" values and are dependent on corporate-produced goods will respect the money we can offer them for their land and the minerals beneath it. They are easily manipulated and exploited. Let them build a casino, the exploiters of Native Americans say, and they'll adopt our values, join our culture, and stop bothering us.

This situation bodes ill for the survival of tribalism. The history of the Bureau of Indian Affairs (BIA) shows that rather than helping tribal people retain their identity and culture, the BIA works to assimilate them into ours so we can exploit them as laborers or take their property away from them. Even the apparently beneficial BIA programs such as Indian schools end up resulting in cultural destruction.

A core concept of tribalism is respect for other tribes. This doesn't mean that you have to like them, or even be friendly toward them. But you respect their uniqueness, their traditions, and their right to their own way of life. And you don't convert them to yours.

When we see that it is wrong to impose our products, way of life, religion, and anything else *simply because it is disrespectful of their way of life,* then this type of Younger Culture domination of one people by another can stop.

Oscar Wilde once noted, "As long as war is regarded as wicked, it will always have its fascination. When it is looked upon as vulgar, it will cease to be popular." Similarly, when we see that the forms of economic and cultural imperialism our culture practices in the name of "free trade" and "modernization" are vulgar and destructive to the health of the human community, perhaps we'll be able to stop the ongoing exploitation and destruction of the resources and people of the Third World.

The line should be drawn at the very idea of imposing anything on anybody. Others have a right to their lands and their lifestyle, no matter how bizarre or dysfunctional these may seem to us. It follows that we have a right to our lifestyle, as odd as it may seem to them. When one or the other is based on domination and conquest, however, or engages in enforced evangelism, this violates the boundary lines.

Respecting the Sabbath for the land and Jubilee

The phrase "sustainable growth" is popular right now among groups ranging from environmentalists to big business. Growth is possible so long as there is something to grow into, and resources to use to fuel that growth. But what happens when we hit the limits of what we can grow into? How is growth then possible?

World Bank economist Herman Daly, in his book *Beyond Growth,* examines this question in considerable and thoughtful detail.[2] He points out that when nations or the world population as a whole reach a point where they've "filled up" most of the places into which populations can grow, and the fuel sources and other natural resources are becoming scarce, "sustainable growth" is no longer an option—but *sustainable living* becomes critical.

While most city-state civilizations in history have self-destructed, at least one we know of had built into it a system of checks and balances that could help overcome some of the structural problems built into the city-state model. While nobody still worships the gods of the Sumerians, Greeks, or Romans—their civilizations are gone—the deity of the Hebrews is still worshipped, in one way or another, by members of three of the world's major religions. And it was the Hebrews, the ancient Jews, who had built into their code of civilization a set of checks and balances.

The Jews called these systems the Sabbath and the Jubilee. Most people know of the Sabbath day—the day of rest. But in the Old Testament the concept was carried much further. Every seventh year the land was required to have a Sabbath, too, and no crops could be grown in that year. (This practice and other parts of this are today practiced by some in Israel.) This provided a rest for the land so it could recover its fertility, and provided a basis for "sustainable" agriculture thousands of years ago.

Another problem with city-state structure is the excessive accumulation of wealth among the ruling or merchant classes. Daly suggests that when the difference between the wealth of the richest in a society and the poorest in a society exceed something between 10:1 and 20:1, then the society will become unstable. We see the 10:1 ratio built into the U.S. military and the civil service—the most highly paid general in the army makes about 10 times what a buck private earns. In universities it's similar, although the ratio often goes as high as 20:1 between the president of the university and the lowest-paid janitor.

After the American experience of the Robber Barons amassing

mind-boggling wealth (and its accompanying political power) in the nineteenth and twentieth centuries, our tax system was organized to discourage such huge inequalities from springing up. In the first half of the twentieth century, the highest tax bracket in the United States was around 90 percent. This turned out to be a social stabilizer in many ways, although that stability has broken down since the Reagan era as the rich have gotten vastly richer and the poor vastly poorer.[3]

For example, according to statistics compiled by Jeff Gates and the Shared Capitalism Institute, "The wealth of the Forbes 400 richest Americans grew an average 1.44 billion each from 1997 to 2000, for a daily increase in wealth of $1,920,000 per person. The financial wealth of the top 1 percent now exceeds the combined household financial wealth of the bottom 95 percent. The share of the nation's after-tax income received by the top 1 percent nearly doubled from 1979 to 1997. By 1998, the top-earning 1 percent has as much income as the 100 million Americans with the lowest earnings. The top fifth of U.S. households now claim 49.2 percent of the national income while the bottom fifth gets by on 3.6 percent."[4]

The ancient Hebrews, however, had a system that restored social stability periodically without resorting to taxation (which often only encourages the unrestrained growth of government). It was called the Jubilee, and occurred on the year following every seventh cycle of seven years. Every 50 years, all debt was forgiven, slaves were freed, and the distribution of wealth was equalized. The 49-year cycle of accumulation and growth was spread among the community, and the result was the only tribe that began just after the times of Gilgamesh and endures to this day.

There were other Levitical systems to keep the tribal flavor in the city-state structure of the Jewish nation: the king couldn't accumulate excessive wealth, widows and orphans were cared for by the community, and earnings could come from "work" but not from "capital"— it was against the law to charge interest.

While it's unlikely that any of these systems will any day soon be imposed on our world or nation, they do provide fascinating food for thought for those who are considering creating intentional communities. They have survived the test of time, and may also be useful seeds if the time ever comes when our society breaks down to the point where serious structural reconstruction is necessary.

Older Culture wealth

In tribal societies the concept of wealth—as we know it—is almost unknown. The night before writing these words, I attended an advance screening of Geoffrey O'Connor's 1997 documentary, *Amazon Journal,* and a talk by him at a bookstore in Santa Monica. In that documentary, O'Connor revealed numerous Brazilian government officials, gold miners, and others rationalizing their invasion of the lands of the Amazon rainforest's Kayapo Indians by saying that the Indians were "primitive" and "poor." This is the line of thinking that says, in essence, we *help* people when we take the gold and wood from their lands and, in return, give them food, guns, televisions, and other things they need to "become modern."

However, the Kayapo tribal economy—at least before the arrival of the white men—had a different basis from ours. Instead of wealth being a function of who had the most things or who controlled the basics of survival such as food, wealth was a function of the entire tribe's ability to support itself as a single entity, and to facilitate for each member of the tribe their own personal opportunity to daily touch the sacred presence of the Creator through the creation. Each person's role in the tribe is to support every other person. Instead of producing the wealth of goods, the tribe produces for its members the wealth of security, safety, and a context for experiencing the sacred.

At first glance, safety and security may not seem like a big deal to the average middle-class American or citizen of western Europe, but half the planet's population who struggle to live on the equivalent of less than $2 per day understands its significance well. Even middle-class Americans have a below-the-consciousness understanding of this: we crave security and safety. As Abraham Maslow, the founder of humanistic psychology, pointed out in the 1950s, security and safety are fundamental human needs. Once these are met, people will strive to reach the highest human need, the need to experience the sacred, what Maslow called self-actualization. Maslow also noted, however, that the majority of people in our modern world live in such a state of perpetual insecurity that they never reach (or even reach *for*) self-actualization.

But security, safety, and the context they provide for daily contact with the sacred are invisible. When Younger Culture invaders encounter Older Culture tribal people, they see only the lack of goods, and therefore assume that the people are "poor" or "living in the Stone Age" or "primitive." In fact, in the commodities that are most essential to creating spiritually, mentally, emotionally, and even physically healthy humans, most Older Cultures are rich beyond the imaginings of Younger Culture citizens.[5]

This, then, is another way to re-create and build community that sustains itself and works: make the primary job of the community to provide safety and security for all of its citizens, and a context in which they can daily touch the sacred.

Just as most tribal cultures have done for over 100,000 years. . . .

Renounce the Destruction of Life

 In the arts of life man invents nothing; but in the arts of death he outdoes Nature herself, and produces by chemistry and machinery all the slaughter of plague, pestilence, and famine.
—GEORGE BERNARD SHAW (1856–1950)

Our dominator culture is, in many ways, a cult of death. Our leaders and image-makers seem to love war. They use the term to describe actions we consider good, such as the "war on poverty" or the "war on illiteracy" or the "war on drugs." Ironically, our war on insects has brought us an actual increase in insect-caused crop losses over the past 40 years. Our antibiotic war on germs has brought us new and incredibly virulent forms of easily transmitted, common, and now lethal bacteria. And, of course, human warfare has caused death and destruction to generations since the beginning of our warrior civilizations seven thousand years ago.

Oscar Wilde was absolutely correct, in my opinion: war is vulgar. The continual glorification of this killing through nationalism and media and dominant culture in general only guarantees more pain and human suffering in the future. The mythos of the warrior-hero is intrinsic to industrialized culture East and West. This mythos enabled Hitler to gain the support of his people as he moved against neighboring states. This enabled Tojo to do the same in his war against China. This cultural myth ensured that "pioneers" who "conquered the West" would be viewed romantically by Americans (and other nationalities).

One cannot fight against war: one can only see it for the vulgarity it is and choose, as the Shoshone people did for ten thousand years, to walk away from it.

The idea that everything in the world is for our use and the war mythos are inextricably intertwined. If another non-human life-form begins to compete with our Younger Culture for our food or the space to grow that food, we exterminate it.

On the other side of human culture are people who see other life-forms as having the same right to the Earth as do humans. These peoples' cultures are usually organized in the tribal/cooperative/community fashion discussed earlier, and just as they work to cooperate with other humans, they also cooperate with nature. While they may (and do) compete with other species for food, they do not destroy those competitors. As Daniel Quinn so elegantly points out in *Ishmael,* the concept of "you may compete but you may not destroy your competitors" is one of the basic laws of nature.[1] With very few exceptions, animals and plants compete with each other for food and access to sunlight energy, but they do not set out to destroy other species as part of that competition.

This concept of competition—as an alternative to genocidal warfare—is one that we need to incorporate into the weave of our cultural fabric. Doing so begins with a critical mass of individuals seeing and understanding its importance, and sharing that story with others.

In small ways we can begin this—organic gardening, for example, competing with the insects and weeds for our food but not exterminating them. In larger ways, we can do this economically by doing commerce as much as possible with local vendors, building the local community.

I was at a talk Bill McKibben gave at Middlebury College in Vermont, and he told the new students how he lives a "one store" policy: if there is more than one of a store, he doesn't patronize it. What an elegant statement of the renunciation of the type of economic "take no enemies" warfare that has laid waste to small, family-operated businesses and local economies across the world! To the extent I can, I have been following his lead: I suggest you consider it, too.

And, in the largest ways, we can work to create cooperative businesses and communities that operate on this principle, to infiltrate the concept into government, to spread the idea through our writing and speaking.

Look into the Face of God

After you have exhausted what there is in business, politics, conviviality, and so on—have found that none of these finally satisfy, or permanently wear—what remains? Nature remains.

—WALT WHITMAN (1819–1892)

One of the first rules of tribal life is that humans are dependent upon their environment. When city-states were created, the man-made environments of buildings and streets helped us forget our sacred connection with the Earth and all living creatures. We viewed our sustenance as coming from the artificial environment of the city, and so our places of worship moved from the cathedral of nature into man-made buildings. Eventually our spiritual disconnection from nature became so profound that Younger Cultures came to view the natural world as evil or "pagan." For centuries, those who practiced outdoor worship were hunted down and put to death by Jews, Christians, Muslims, Hindus, and other Younger Cultures and religions.

Now our environmentally destructive actions are bringing us face-to-face with the natural world we have so long neglected or treated like an enemy to be subdued. It's time for us to recognize other living things as our equals in their claim to life on this planet. They are our older brothers and sisters in the history of the Earth; they are inextricably connected with us and our life source.

My mentor Gottfried Müller once said to me, "Thomas, do you want to know how to look into the eyes of God?"

"Of course!" I answered.

"Then look into the eyes of any other living thing," he said. "There in the eyes of a cat or dog, in the eyes of a fly or fish, in the eyes of a friend or enemy, you are looking into the eyes of God."

On another occasion, he told me, "My teacher Abram Poljak often said that if you bless one blade of grass, then all grass will bless you. If you bless one tree, then all trees will bless you. *But to bless is also to thank and respect and love, it is not just to say, 'Bless you.'*"

It's not just human societies who have a right to their own unique existence, mandated by the natural law that diversity strengthens living systems: *it's true of everything on the Earth*. For example, if loggers were to see trees as sacred living things, then large-scale clear-cutting and non-renewable destruction of old-growth forests would become not an "unfortunate necessity" but instead a blasphemous obscenity.

For the first 194,000 years of humanity's 200,000-year history, humans viewed the world and its living creatures as sacred, as having souls or spirits. A person who caused permanent harm to that world was condemned as insane and banished from the tribe. The members of the tribe realized that he was destroying the world of his children's children, an unthinkable and aberrant act. The ancient peoples understood: when you kick your Mother (Earth), she kicks back. She does not roll over and submit to her own death.

Does the world consider human life more important than a tree or a fox? Does the forest "prefer" men to deer? Do the oceans thrive as a result of our presence? Is the planet healthier for our having inhabited it these past seven thousand years since the rise of the city-state?

Only the arrogance of a Younger Culture could come up with the idea that all of this planet's and this universe's history is directed toward one's own lifetime and nothing beyond it. Because we were able to steal/kill/conquer it, we concluded, some god was on our side and had, in fact, decided even before the fact that it should be this way. Instead, we must see all life as sacred, as our distant ancestors all did.

You can share this idea with others. Tell them about it—suggest to a friend that when he or she is looking into the eyes of any other living thing, s/he is looking into the eyes of the Creator. What a startling concept—and what a powerful one. That person may be changed by the experience, and, in turn, share it with another.

For some people, just connecting to or seeing divinity or the possibility of the presence of God in daily life isn't enough: they have to *understand* it. I know—from my own personal experience—that when an intellectually driven person *understands* that the Older Culture view of the world was more scientifically valid than ours, then that person is transformed. That change begins with our understanding of "how things work." Out of that understanding, our idea that all creation is part of us and us part of it—and therefore it's all sacred and has value—grows to certainty.

Change the Focus of How We Use Technology

We cannot cheat on DNA. We cannot get round photosynthesis. We cannot say, "I am not going to give a damn about phytoplankton." All these tiny mechanisms provide the preconditions of our planetary life. To say we do not care is to say in the most literal sense that "we choose death."

—BARBARA WARD (1914–1981), *Who Speaks for Earth?*

In the opening chapters of this book, I pointed out how our lifestyle and, indeed, our entire worldwide modern civilization are possible only because we're rapidly using up a 300-million-year-old non-renewable resource: ancient sunlight, principally in the form of oil, but also coal and gas. I also cited figures that indicated that this resource—at current rates of consumption—will run out in our or our children's lifetimes.

The way it'll most likely play out, though, is nowhere near that simple. We won't just one day suddenly wake up to a world of dry gas pumps and grounded jetliners. Instead, as oil becomes progressively less available, its price will rise. This rise in price will affect the price of everything made from or with oil—from plastics to manufactured goods to the food we eat produced by oil-powered farm machinery and transported in oil-powered trucks and trains.

As it did during the oil crisis of the early 1970s when oil prices temporarily shot up, this will produce economic crises, exacerbate the gap between rich and poor, and stress the social fabric of countries worldwide. A return of conditions such as prevailed during the Great Depression is not inconceivable and, given that the world now has three times more people on it than it did in 1930, the situation may even be far worse than it was at that time. Some futurists are predicting "oil wars" and global conflicts over the ownership of energy resources.

Whatever the details of the way increasing oil scarcity will affect the world, one thing is certain: people will be forced to use less oil. Because of this, the 40-years-and-we're-out prediction is unrealistic. Instead, sometime in the next decade or two, as oil wells begin to run dry around the world or countries decide to hoard the reserves they still have, rising oil prices will force consumers and nations into less oil-intensive ways of living.

Use our oil to not use oil

While we still have a chance, let's use what energy resources we have to develop renewable alternatives.

Oil currently fires the furnace of industry and government. But we're using it in a "once-through" fashion—we burn it, and that's that: the resource is gone, never to produce another benefit. That's precisely the kind of mistake Dwight Eisenhower meant when he said that the building of war machinery represented a theft from our children: he was referring to the "once-through" nature of military spending. If the government uses tax dollars to build a bullet (or tank or missile), there is a short-term stimulation of the economy as the result of that expenditure. Somebody is hired to manufacture the bullet, somebody else mined and smelted the lead, and so on.

Over the short term, it stimulates the economy as it increases employment and consumes materials extracted, refined, and manufactured by industry. We've seen this in the short-term economic benefits of military spending during World War II, the Korean War, the Vietnam War, and the trillions of dollars spent during the Reagan administration on the Star Wars program.

The problem, however, is that once that money is spent and those jobs are finished, we hit a wall. The only thing military armaments can be used for (economically, not politically) is their own destruction. When a bullet is fired, it's gone. No more use can be put to the gunpowder that burned or the lead slug that's now buried in somebody's body. When a tank, bomber, or a missile is used, there is no secondary gain to the overall economy. (Of course, there is the ripple effect through the economy of the workers who built the bomb spending their pay, but that is minor compared to the multiple ripples that would happen if the bomb itself were "productive.")

On the other hand, when the same money and resources are used to build a commercial long-haul truck, that truck participates in the economy for its useful life span, facilitating commerce and contributing value to the economy every day. It's no longer a once-through expenditure; it's become something that works with the rest of *the system* to produce more value. Building a bomber is a onetime expenditure, as if the money were poured into a hole and buried; building a commercial jetliner creates an economic tool that then can provide employment and transportation for thousands of people for decades.

Particularly important are the products that, when used, capture current sunlight energy and transform it into a form that can replace fossil fuels. Such products have an ongoing useful and productive life, and they also reduce the future amounts of fossil fuels we'll need.

As such, they could be viewed as putting capital into our energy bank, rather than simply removing energy from it. Solar panels, wind power systems, hydropower systems, hydrogen production and storage systems: all of these represent ways that current oil can be used as an investment rather than an expenditure.

The ultimate resource

A few decades ago, the late Julian Simon (a darling of the Republicans and the World Trade Organization) suggested that there could never be a global resource crisis because, he said, humans are so terribly inventive that no matter what happens, we'll just think of a way out of it. On one level it's the sort of naive thinking that led the Maori people of New Zealand to cannibalism when they ate the last of the Moa birds, and brought civilizations from the Sumerians to the Romans crashing down. On the other hand, it contains a germ of truth.

While we're not smart enough to have avoided fouling (and over-populating) our nest already, our ability to develop solutions is probably far better than we can imagine. The May 2003 issue of *Discover* magazine, for example, had a feature article titled "Anything into Oil" that documented a new invention that can turn sewage sludge, agricultural waste, and even plastic and paper from a landfill into high-quality oil. Fuel cell development is dramatic, and the possibility of a hydrogen economy that's driven by renewable sources (rather than the hydrogen-extracted-from-oil favored by Big Oil) is being seen increasingly as a very real possibility.

There's also the issue of cutting consumption and reducing waste, which could quickly change the ratios of resources to humans on the planet. For example, according to a 2002 report in *Environmental Science & Technology*, it takes 72 grams of chemicals, 1.6 kilograms of fossil fuel, and 32 kilograms of water to manufacture a single 2-gram computer microchip. Extrapolating from that, and adding in transportation and power consumption overhead, it may take between hundreds of pounds and a ton of raw materials to produce a single personal computer. Using it for an extra few years (or an automobile, which is similarly resource intensive and has its own assortment of microchips) would, if widely done, significantly reduce the load on the biosphere.

If we as a society begin to use our fossil-fuel resources wisely to wean ourselves off the need to use fossil fuels for heat and electricity production, then the impact of "the end of oil" can be softened. At the same time, we'd be reducing our consumption of oil as these alternative-energy systems become available. While ultimately this will have to happen nation- and worldwide, it's already beginning on a small-scale basis in homes and rural communities all over the world.

Here in Vermont, electricity is pretty cheap. It costs us about nine cents for a thousand watts of electricity for an hour, so having ten 100-watt bulbs fully illuminating different rooms in the house at the same time costs less than a dime an hour. But that can't last very much longer.

Living "off the grid"

There's a growing movement in the United States to generate one's own power. It started a few decades ago, mostly by people living in remote areas where bringing in power from the local utility grid was impractical or costly. Over the past 20 years or so, with the development of efficient and inexpensive wind, water, and solar generators that are practical for home use, it has spread to people who value their independence, have concerns about the reliability or cost of future electric supplies, or are cautious about the ecological impact of "big electric."

It's now technologically possible for most suburban and rural dwellers in the industrialized world to generate their own electricity. Sanyo of Japan manufactures roofing tiles and window panes that are solar-electric generators, and in many parts of the world roof- or yard-mounted wind generators can power a home. The cost of solar-cell-produced electricity has dropped from over $30/kilowatt-hour in 1975 to less than 30 cents per kilowatt-hour in 1996, a 100-fold decrease that dropped again to 7 cents per kilowatt-hour over the next five years. Storage batteries and inverters are dropping in price, and the hydrogen-powered fuel cell (currently used only by astronauts) holds great promise for power storage, since running electric current through water can easily produce hydrogen.

Similarly, most homes could grow their own food in a pinch. An acre of prime land can produce 50,000 pounds of tomatoes or 40,000 pounds of potatoes in one year. In many parts of the world (particularly in many small European towns), it's fashionable to turn the front lawn or backyard (or both) from grass production into a huge vegetable garden, which often supplies a significant portion of the family's food needs. Many Americans remember that this was common in this country during the Depression and up until the end of World War II (these plots were referred to as Victory Gardens).

Water purification systems have come a long way, too, with hand-powered reverse-osmosis filters capable of detoxifying rain- and groundwater almost anywhere.

The idea of moving "off the grid" is a popular one in rural areas and among those who see the government as a malevolent force. As such, it's very much a minority way of life. However, decentralization of power, food, and water production may well hold one key to how we can come through the coming changes in world oil availability without collapsing into chaos and tragedy. This holds promise because according to a 1990 U.S. government study, renewable energy sources (solar, wind, water, biomass) could supply over 70 percent of the power requirements of this nation. In California alone, for example, over 15,000 wind generators now produce enough electricity to theoretically light the city of San Francisco.

Government subsidies for the production of energy, however, are largely limited to the huge oil- and coal-based power generating companies with enough campaign donation dollars to sway legislation. Concerned about future generations' dependence on dwindling oil supplies, Jimmy Carter introduced subsidies for small-scale electricity production, which jump-started an industry, but under pressure from big-oil campaign contributors Ronald Reagan eliminated them as one of his first acts of office, causing the embryonic small-scale solar industry to die a sudden death. Nonetheless, a small remnant of that industry has struggled back to life, and increasingly people are experimenting with small-scale solar, wind, and hydropower.

While large-scale centralization may seem economical, ultimately it's not. Centralized, hierarchical structures are inherently less stable than decentralized, grass roots ones. Monolithic systems richly benefit those who control them, but offer only ongoing dependence to their customers.

In a story reminiscent of how American companies are removing natural resources from Third World countries and then selling finished goods back to those countries (and controlling agribusiness while wiping out family farms), Mahatma Gandhi used pictures of the simple spinning wheel, a handmade tool to convert wool or cotton into thread, as the symbol of his nationalist movement against the British. At that time, the British had ordered the shutdown of all clothing manufacturing facilities in India, and shipped cheap Indian cotton to England to be made into clothing by British workers. While this provided work for the English citizenry—something pop-

ular in the English countryside, hugely profitable for the owners of
the clothing factories, and politically helpful to the British government—it impoverished the Indians, who were forced to pay high
prices for clothing imported from England which they, themselves,
had been inexpensively manufacturing only years earlier.

Gandhi argued for a return of local economies rather than centralized ones, and suggested that families or, at the largest end of the
scale of practicality, villages should grow their own cotton, spin their
own thread, and make their own clothing. He did this himself, making his own simple clothes by hand, and soon the logo of the spinning wheel was a powerful emblem of change all across India, as well
as the unofficial logo of his independence movement.

As Gandhi well knew, when people produce their own food,
heat, and light, they are more free and independent. Even more
important, they are usually more efficient in their use of these
resources, because they're so familiar with them. Looking at their
own light, eating their own food, and feeling their own heat, they
have an intimate knowledge of the significance and importance
of these essentials to human life that many people living "on the
grid" lack.

Conservation

When we moved to Vermont, we quickly discovered one of the
unique aspects of living out in the country on the side of a mountain:
power outages. In our first month here, we had three days with no
power. The locals say that it's not generally this bad—the weather had
been unusually severe—but nonetheless we quickly learned how to
use an emergency generator, to light a house with candles and oil
lamps, and to appreciate battery-operated radios and computers.
Which led me to discover how really wasteful my consumption of
electricity is—and how relatively easy conservation is.

Since conservation reduces the amount of electricity we need to
generate, it has the further benefit of making it easier to live off the
grid. Take lighting, for example. It's only in the past hundred years or
so that we've had the idea that an entire room need be lit for occu-
pancy. For all the rest of human history, we used "area lighting": a

whale or vegetable oil lamp to read by, or a beeswax candle for conversation. These forms of lighting consumed insignificant amounts of fuel—the equivalent of a 10- or 20-watt bulb, at best.

Similarly, many people are discovering that it feels very satisfying to live efficiently. Driving a bicycle instead of a car; saving and reusing food packages; recycling table scraps into a compost pile; buying secondhand clothes and repairing the old ones we've kept; super-insulating the house so it uses less fuel; maintaining the car so it can reach 200,000 miles and still run well. There's a sense of accomplishment in living frugally, a feeling of independence. In recent years, frugality has even been touted in consumer and women's magazines as a fashionable way to live.

Nonetheless, there's a nagging voice in the back of many people's heads, perhaps an echo of the Reagan years, that to voluntarily not consume, to not grow and compete and acquire and dominate, is somehow an admission of failure. Could it be?

On the contrary, it's an act of self-preservation and qualifies as highly successful behavior.

Turn Off the TV

If we could sniff or swallow something that would, for five or six hours each day, abolish our solitude as individuals, atone us with our fellows in a glowing exaltation of affection and make life in all its aspects seem not only worth living, but divinely beautiful and significant, and if this heavenly, world-transfiguring drug were of such a kind that we could wake up next morning with a clear head and an undamaged constitution—then, it seems to me, all our problems (and not merely the one small problem of discovering a novel pleasure) would be wholly solved and earth would become paradise.

—ALDOUS HUXLEY (1894–1963),
"Wanted, a New Pleasure"
(1949 essay)

Something has arrested the social development of our society, freezing most people—almost all of modern civilization—in an immature stage of awareness about life. It appears that this is a root cause of why today's humans are capable of ruining their environment for short-term gain *even when most people know they're damaging the future* for themselves and their children.

It's a matter of being connected to life and of maturity.

Spending time with Native Americans and tribal people in other parts of the world, I've found people who live in a way that is *daily and continuously meaningful.* This was an extraordinary insight: a shock, really, to me, since I'd been brought up in a culture that told me that our way of life was the best, the most free, and the happiest.

From my contacts with people of Older Cultures, I've become convinced that our sense of spiritual disconnection started with our Younger Culture's disconnection from nature. (One metaphor for this was the expulsion of Adam and Eve from the Garden of Eden.) When we decided to separate humanity from all of the rest of creation, we created a schism that was deep and profound. When we decided that the world was *here for us, separate from us,* and *it was our holy duty to control and dominate it,* we lost touch with the very power and spirit that gave birth to us.

So we see people who are spiritually disconnected, living in boxes and driving in boxes, perhaps once a year going "out into nature" to get a brief taste of what was once our daily experience. These people seek escape. They sit in urban and suburban homes and feel miserable, not knowing why, experiencing anxiety and fear and pain that cannot be softened by drugs, TV, or therapy because they are afflicted with a sickness of the soul, not of the mind. They've become disconnected.

This disconnection has caused the world's dominant culture to become stuck in a stage of arrested development, more concerned with "me" and "now" than with being custodians of the future, in contrast to the way Older Culture peoples view the goal of day-to-day life and, indeed, the very meaning of life itself.

As people grow from infancy to adulthood, they go through many stages. Babies are born self-centered, unaware of anything outside their own instantaneous experience. At certain ages they become aware of their surroundings and their role in their surroundings. Gradually babies realize that something can exist even if it's invisible, like a toy that's under a blanket or a parent who's out of the room. Children of a certain age become able to care for themselves instead of depending on somebody else to do everything for them. Teenagers become conscious of social structure among their peers, become able to carry the responsibility of baby-sitting, and so on.

Part of the maturing process is to shift away from a "me" orientation to a sense of "us"—a responsibility for others.

This was no problem at all in Older Cultures: traditions consistently reinforced an individual's interrelationship with their environment and their responsibility both to and for their community.

But our Younger Culture has two problems here. First, when we decide what we're responsible for, we often draw a narrow circle around ourselves, saying, "I'm responsible only for the people I'm directly involved with. Everybody else is on their own. Somebody somewhere is responsible for the planet—not me." With maturity, we're supposed to grow into a more effective view of the world, but the evidence is that in Younger Cultures this view usually doesn't materialize.

The other problem is even worse: our Younger Culture increasingly gets stuck in an unfinished, immature stage of development, a stage that says, "I'm the center of the universe, I'm what's important." At this stage of life there's not even any question in the child's mind about providing for the future; it's "Gimme now."

Why does it seem that our culture is becoming progressively less mature, rather than more mature?

The primary "immature" cultural concept—"You are the most important person in the world"—is shouted at us daily through TV, the primary spokesvehicle of our culture. The constant reinforcement of this message keeps our culture immature and prevents us from growing in maturity.

The reason for the persistence and intensity of these messages is simple: when people behave like children, wanting immediate gratification for their every desire, they are the ideal consumers. Only when we turn off the messages can we begin the process of maturing—and this is rarely done.

The things that catch our eye, catch our ear, are things that excite us. That's no surprise, and it didn't take a malicious advertiser to realize it. As advertising became more and more skilled, more and more capable, people found more and more excitement and stimulation from what the ads offered. "Look what you can have," the ads say: "Look how your life can be better or more fun."

"Consumerism" was born—the idea of making your life better by getting something—and amplified by television into a culture-consuming monster. It works well for corporations and marketers in a "me-centered" world, but it distracts humans from the more mature view of making life better *for everyone.*

Consumerism has largely subsumed even the churches, converting their holy days into orgies of consumption, and today far more people pursue happiness in buying new "things" than engaging in the rituals of any organized religion.

A Stanford University study of 1,533 ninth-graders found that youngsters who rented movies instead of watching commercial television were less likely to start drinking alcohol while in school. Just one hour of television a day produced a 31 percent greater risk that in the next 18 months a child would begin to drink, and every additional hour increased that risk by another 9 percent.[1]

While I'm not arguing against participating in the modern-day marketplace, what I have seen over and over again is a qualitative change in both adults and children when they remove television from their homes. The home quietens, the people become more centered and anchored, the real world is noticed.

Turn off the TV, and sit quietly for just 10 or 15 minutes each day. Take another few minutes a day to walk outdoors. Your life will change for the better, and in that way you contribute to the healing of our planet.

The Modern-Day Tribe:
Intentional Community

 The community stagnates without the impulse of the individual. The impulse dies away without the sympathy of the community.

— WILLIAM JAMES (1842–1910)

The most basic unit of the tribe is the family. Many tribes in places like the Kalahari desert are comprised of a single extended family, often only a dozen or so people. In Western society we see some families that have organized themselves into something resembling a tribe: the Kennedys are a well-known American example. At least there's a sense of "this is us" among them. Seeing your family as a tribe helps build identity, strength, resilience, and the critically important sense of belonging.

Many people today have chosen to sacrifice family culture (in which it's important to be together, to build a sense of shared experience) for the culture of consumerism. Both parents work so they can afford a larger house and a newer car, when a smaller house and car might have allowed them to have more time at home with each other and the children. They sacrifice important family bond-building rituals, such as dining together or long family discussions, for hours of corporate-funded and advertising-driven television viewing.

Increasingly, however, people are choosing to simplify their lives, reduce the time they spend in the business world (and its primary promoter, television), and redirect that energy to their own family—in a very real sense, their own tribe.

It is possible to step back from the television/consumer/mass-media feeding frenzy of "more, more, more!" and modify one's consciousness and behavior. One way is to get everybody in the family involved in a more direct way in the daily necessities of living. This allows people to feel and see the connection between their efforts and the food they're eating or the heat they're enjoying.

There is a growing trend among conscious people toward moving to places where there is enough dirt around the house to grow some of your own food, and enough trees to renewably harvest wood for heating. You can find a local source of drinkable water and look into generating your own electricity, build a root-cellar and put up food for the winter. Living more delicately on the Earth, many people are now choosing to live as a local family tribe.

Intentional communities

The next step up from family tribes is tribes that share a common interest.

These are made up of people who live together in the same area, share their food supply, work together to build their fuel supply, and cooperate in creating living spaces. We see examples of these second-level tribes in the lives of Native Americans, and also in the stories of the Europeans who settled the American West. They created communities of shared survival interest, and, therefore, achieved some measure of security and stability.

In addition to survival concerns, community has two important aspects: safety and human interaction. These have been the foundations of community since the first pre-humans built fires and huddled around them, hoping that the magical warmth would keep them from evil spirits and repel the tigers or bears.

Years ago Sigmund Freud pointed out that any collection of people, be it in a company, a community, or even a nation, would eventually reassemble itself into a family-like structure. Parental figures emerge, sibling rivalries come out, and people find niches and roles for themselves in the community that often match those they had in their own family: the bad boy, the cute child, the little professor, the nurturer, the instigator, etc. Freud went on to show how when the parental figures were dysfunctional (as happened in Hitler's Germany, for example), the entire "family" (the nation) becomes dysfunctional.

Realizing that this family re-creating happens in all communities, regardless of size, an important first step in community-building is to create a nurturing, safe, and functional community structure. Humans have experimented with this structure for a 100,000 years or more. What has evolved as the most highly functional, stable, and sustainable form of human organization is the decentralized, small-tribe, egalitarian, and democratic community structure as practiced by the Shoshone and other native peoples worldwide.

By carefully choosing leaders who have demonstrated their ability to nurture others—who can refrain from beating their own drums or following neurotic agendas, but can give and give and give without getting much back in return—tribal communities provide both the "parent" models for the community, and the sense of safety that people need in order to allow themselves to immerse themselves in the community without hesitation or fear.

These elders are at the core of community: they carry its accumulated wisdom. They are leaders in particular areas because they've demonstrated expertise or experience or wisdom in those areas. By losing our elders (often to nursing homes), and particularly by replacing interactions with them with television, we've lost an incredible store of knowledge. By reinventing our communities in ways that value the wisdom of experience, we can recapture this.

When members of the community feel that the community is a sham, that it's there to serve some purpose that's not congruent with their needs (as in to sell them something, to control them, or exploit them somehow), then community also breaks down. Community members lose their commitment to keeping the streets clean and the walls painted, or start misbehaving in destructive ways. We see the unfortunate result of this in our big cities, but it's just as true of many small towns or communes, particularly those where traditional community interactions have been replaced by people sitting in private rooms staring for hours at TV.

However, small communities are becoming a significant presence in the world. Those who live "off the grid" are increasing in number as well, and these people may have even more impact because they embody and emit the values of independence and self-sufficiency. By their existence, they put these ideas into the ether, into the collective unconscious. They become independent from the power of the corporate and political interests who migrated into the roles of rulers, definers, and reflectors of our culture. And they are separate from the big cities that most clearly personify the spiritual disconnection of our city-state culture.

Some people argue that small "tribal" communities are a bad thing because they're sometimes run as cults. Cults are soul-stealers, mind-control entities that discourage individual initiative and seek to control not just the actions but the very thoughts of their members. I agree that such groups are just another form of domination and are to be avoided. A simple rule of thumb is to avoid any person or group who says that the way to divinity, salvation, enlightenment, happiness, freedom, or clarity is available exclusively through them. This is the definition of a cult, and it becomes particularly obvious when a single individual or leader in the organization or community proclaims him- or herself as the "gate to the divine."

But with that one caveat, small, independent, self-sufficient communities have the greatest ability to survive the normal cycles of boom-and-bust that our economy and culture go through, and an even better chance of surviving the major catastrophes that may loom ahead as our oil supply dwindles. Rural communities from the Utah Mormons to the Amish of Pennsylvania (regardless of how you view their religious philosophies) have weathered the Depression, the wars, and the vagaries of weather and economy due to their sharing of resources and their united front.

Get support and information from the growing community movement

Many people are reinventing community today, in a movement that's grown dramatically since the 1980s. They share ownership of the land with a few other families and create a small tribal community. Some even give themselves a tribal name, and clarify their purpose, intentions, and the ways they are both similar to and different from other tribes.

Not surprisingly, many people in the movement share freely what they've learned. This can provide a real head start for people who want to begin the transition into community life now. From books, mentors, and in classes offered by communities around the world, you can learn pioneer and native skills for growing and finding food in the forest and fields.

Contemporary books with titles like Howard Ruff's *Famine and Survival in America* do have an effect in raising consciousness about these issues, although in some cases they're so shrill in tone that they put people off.[1] In any case, they're usually not read by the mainstream: many—most—people are so occupied with enjoying their current standard of living that they can imagine no other kind of life. They love their cities, and don't stop to think about their dependence on the trucks that bring food to the supermarket. They see their neighbors as a source of stimulation and entertainment, not the potential competitors for scarce resources they would turn into if the trucks stopped rolling and the cities turned into the macabre jungles of futuristic movies. Their view of the world is correct for today, at least in the wealthy cities of the Western world, and no doubt it will continue for some time into the future. The past is littered, after all, with the corpses of those who said the end of the world was just around the corner and never lived to see that day.

On the other hand, many people are re-examining the possibility of independent, small-community lifestyles that are nearly tribal, as much to make good times better as to prepare against bad times. They're forming small cooperatives and moving into the country, sometimes even organizing these groups in the city when they can get enough land to grow food in their yards or a common area. The movement toward community in the world is growing so rapidly it's spawned books, magazines, and directories such as *Creating Community Anywhere*, *Communities* magazine, and *Communities Directory: A Guide to Cooperative Living*.[2]

It's popular to deride "utopian communities," particularly in scholarly works on the environment, but such communities have a history of succeeding—and surviving difficult times particularly well. From small-town early America to communes scattered from Vermont to California to Bali, from the kibbutzim of Israel to the indigenous societies of the world, small tribally organized societies do work.

A visit to an "intentional community"

One modern-day version of these are called intentional communities. Often they're organized as a small corporation, where everybody owns the land jointly, but individual families own the buildings on them as well as "shares" in the larger community. If somebody wants to leave, they sell their shares and buildings back to the community; if somebody wants to join, they must be checked out and approved by the community as a whole. *The Communities Directory* has an exhaustive listing of such communities around the world, as well as many chapters on how to go about starting your own community.

Reading that directory in the summer of 1997, Louise and I learned of just such a community near us, started decades ago by people who wanted a new kind of life. We contacted them to arrange for a visit, and during Labor Day weekend we went to see for ourselves.

* * *

Quarry Hill is the oldest intentional community in Vermont, having been started 50 years ago originally as a writers' and artists' retreat by Irving and Barbara Fiske, and their shared vision and mission is the nonviolent rearing of children. The sixties brought changes, as did Irving's fascination with the writings of William Blake and other philosophers and mystics, and now there are 26 houses on over 200 acres of breathtakingly beautiful mountain land, as well as a private school where their shared mission is lived out. Louise and I spent the day at Quarry Hill, seeing the land and houses, meeting the people, and hearing their stories.

"One of the things I like the most about this place is the sense of safety," said Judy Geller, who heads up Quarry Hill on-grounds school and has lived in the Vermont commune for 20 years. "People are very accepting here. You can just be yourself. You're safe to be yourself."

While Quarry Hill is the oldest intentional community in Vermont, it's by no means the only one. About a dozen others exist in the state, nationwide there are over 500 listed in *The Communities Directory,* and the publishers of the directory point out that several hundred additional communities asked not to be listed (presumably they're full), so the "known" number is about a thousand in North America.

Communities are by no means identical, and most are not "communes" where varying levels of personal property are owned communally. The basic structure of all communities is a grouping of people who have chosen to live together for a specific purpose. Those purposes range from political or environmental activism to spirituality to music to operating health, teaching, or conference centers.

There's a myth that the contemporary intentional community is a remnant of the hippie lifestyle of the 1960s, and/or that such communities all died out with Reaganomics and Yuppies. But it's just that: a myth. Intentional communities have existed for as long as people have, and in my experience the majority of contemporary intentional communities bear no resemblance whatsoever to the hippie crash pads of old.

A contemporary suburban subdivision, or—for that matter—a neighborhood in a town or city, is a community. People have chosen to live there, usually for reasons of quality of life, status, economic necessity, proximity to family, or convenience. In these types of communities, however, there is no structure that determines who can move in: if you have the money, you can buy a house.

In an intentional community, on the other hand, the shared sense of purpose, beyond simply a place to live, tends to act as a first-level filter, ensuring that only people with congruent goals are applying for membership in the community. While in a contemporary subdivision or urban apartment it's possible to live in the same place for years and never have a meal with your neighbors, such a disconnected co-existence is nearly inconceivable in a well-functioning intentional community.

The downside of this process of communities pre-approving new members is that it's easy for communities to become too homogenous. Even assuming that things like racism are not a problem because of well-intentioned community members, there's still a tendency among communities to unintentionally select for sameness rather than diversity. At Quarry Hill, several residents made a point of telling me how important diversity is to them, and how that diversity is one of the things that keeps their community strong. Residents include whites, blacks, Americans, Germans, and even a New Zealander, as well as people of all ages, from one to over eighty. Keeping this point in mind is probably a good thing for community builders.

Intentional communities usually arise from one of four roots:

Genetics

This is the basis of the traditional tribal community—everybody is a member of an extended family. This is the core type of community found among indigenous people around the world, and has formed the core of community among humans and pre-humans for millions of years. While this type is common among Native peoples (there are over four hundred different Native American tribes in North America, for example), it's rarely the basis of a created-from-scratch intentional community. However, it does naturally happen, particularly over time: the Oneida community in New York State was started in 1848 and is now populated by fourth- and fifth-generation people (among others), and at Quarry Hill residents include the children and grandchildren of the founders and early residents.

One of the advantages of genetic community is a sense of continuity and connectedness with one's ancestors. This leads to respect for elders and the young: the elders carry the knowledge of the past, and the young are the future. I spoke at a conference for Native Americans sponsored by the University of Oklahoma (I was one of only five or six whites among the four-hundred-plus participants and presenters), and during the opening ceremonies the elders were asked to come up front to give their blessing. Forty-four gray-haired men and women came to the front and one led the prayer in his Native language. Then the man who had brought the greatest number of elders to the meeting was honored for having facilitated their presence.

During that conference, every person, when introducing him- or herself, always included with their name their tribal identity (and, usually, in their language, their "Indian name"). These identities are real and solid and meaningful to Native Americans, and often based in some way on the names and behaviors of local animals. In fact, it is in the animals of nature, the ways that they adapt to their environments and have different ways of living, that many of the Native American concepts of identity, purpose, and the essential goodness of "the differences between peoples" are grounded.

In modern America, it is popular to call this concept racist. When Malcolm X talked about people of African ancestry establishing their own sense of identity and their own rituals and culture instead of simply trying to be a part of "white culture," he was branded a racist by many, both black and white. Similarly, those in the "white separatist movement" almost define the term "racist" in their belief that races should live apart and not mix. The reason why "racist" is a negative term, and why this sort of separation of peoples is often disastrous in modern America, is because of our culture of domination, the culture in which it takes place.

In the Native American Older Cultures (cooperator cultures), it's perfectly acceptable (in fact, it's *desirable*) to be tribally/genetically different, and to retain that separate identity, *because that difference does not imply superiority, inferiority, or power . . . but does enhance diversity.* The racists in our contemporary culture, however, coming from the Younger Culture mind-set of "somebody on the top, others on the bottom," present a very real ethical and cultural dilemma for those who want to raise the topic of genetic tribalism. Native American tribalism (called "racism" by some in white culture) works well for them because, in most cases, they are acting out of the Older Culture perspectives of what tribal/racial/ancestral differences mean: different skills, something new to bring to the larger culture, and the idea that people having different traditions, different religious practices, and different ways of living are not only acceptable but a *good* thing.

Our Younger Culture is an absorptive one, eating everything in its path and turning everything and everyone to not just its own use but, in particular, to the use of those who control it. This is why one of its main stories is that "different from us is bad," no matter how nicely our culture tries to dress up that story in phrases like "the great melting pot" and "rainbow culture." The fact of the matter is that America (and much of the rest of the world) is now becoming re-tribalized, and much of that re-tribalization process is happening along racial lines and with tragic consequences.

The reason for this dysfunctional, fragmentary, and often violent re-tribalization is that our Younger Culture long ago destroyed the tribes that blacks, whites, and Hispanics (among others) came from, and so there is no "starting point" for these people to re-create tribal rituals and ceremonies that are meaningful and constructive. The result can be seen most readily among the gangs (tribes) in the inner cities, but is pervasive in our society as a whole.

That people have a need for this sense of "tribal" identity is found in many places, including, in my opinion, in the New Age movement, which is almost exclusively made up of whites who lost their tribal connections thousands of years ago. In Paganism, Wicca, Animism, and dozens of other smaller offshoots, we see attempts (ranging from noble to laughable) to revive Druid, Celtic, Norse, and other ancient rituals and introduce them to the modern-day but now-detribalized ancestors of those tribes. The problem with doing this is that the destruction of the tribes was so complete, there is very little left to begin with. Nobody alive remembers the language of the Celts, their traditions and ways, their creation story, their sacred ways, their rituals. The connection was totally broken, to the point where their holy places were sought out and destroyed by the Roman conquerors.

One of the hot topics at the Native American conference I attended was what Natives should do when approached by whites asking for answers to spiritual questions. I sat in on a workshop where this topic was hotly debated: one side said that the Indians have "an obligation to the world" to share their knowledge about how tribal societies are organized and made to work, and about spirituality that is respectful of the Earth; the other side said that "those murderous Anglos" had already once "done their best to kill us all" and that Native Americans should distance themselves as far away from "those insane whites" as possible "and let them all kill each other, as they tried to do to us."

One speaker, in defense of the idea that Native Americans should help whites learn how to create tribes, said, "Among the whites, only the Jews have any sense of their tribe left. They still have their tribal rituals and holy places and sacred objects, and they still have a sense of bloodline. The rest are lost, and jealous that they have lost their tribal identities, which is why there is so much anti-Semitism among the whites and blacks." It's an interesting, if oversimplified, point to consider: people *need* tribal rituals and a sense of connection to their genetic tribal roots. Those tribes that are Older Culture cooperator-based rather than Younger Culture hierarchical and dominator (there is currently no worldwide Jewish chief rabbi or king, for example), are more likely to survive (even in the face of persecution) without the need to resort to force, violence, or threats of eternal damnation against those who disagree or don't go along.

Nonetheless, the damage is done. I have no more knowledge of the 20,000-year-old tribal ways of my Norwegian, Welsh, and Celtic ancestors than do even the people still living in those once-tribal lands. So if I were to start a community and say in an advertisement seeking members, "Norwegians and Celts wanted," it would be meaningless at best, and racist at worst. Most members of our Younger Culture world who want to try to live out Older Culture ways will have to resort to one of the other traditional ways that community comes about.

Charismatic leadership and a shared vision

Historically, charismatic leadership has been the foundational point of most non-genetically tribal intentional communities: somebody has what others view as a great idea and in addition this person has the charisma and leadership skills to rally others around. In many ways, the beginning of community is often the same as the beginning of a religion: in fact, many contemporary religions began as communities gathered around a strong leader (or the memory of a strong leader, with a strong contemporary "second leader"). Although there may be some religions that didn't start this way, I can't think of any—from the "big" religions like Judaism, Islam, Buddhism, Christianity, etc., to the subsets of these like the Mormons, Methodists, and Shiites. Just as many religions have faded away, particularly the wilder offshoot sects of mainstream religions, so, too, have many communities. Usually the reasons are the same: the leader gives up, moves away, or dies.

Those communities that survive the transition from a strong leader to a strong community are the ones where the leader has allowed power sharing in the community and has left a legacy of a vision people will enthusiastically carry forward into the future.

Commenting on this issue and the spectrum of power and control issues that surround it, Don Calhoun, the husband of Quarry Hill co-founder Barbara Fiske and resident community philosopher, wrote in his book *Spirituality and Community* words that he and Barbara echoed when Louise and I visited Quarry Hill:

> Communities, like all intimate relationships, have to solve the problem of preserving both closeness and individuality. One danger is that the individualistic refugees from social repression who are drawn to voluntary communities will be so bent on "doing their own thing" that the community will fly apart into anarchy. The opposite danger is that the community will be so insistent on total absorption into its goals that personal individuality will cease to exist.[3]

The solution that Don and others over the years have proposed is one of shared vision, wherein the individual is both larger and smaller than the community, and each works to serve the other in an organic balance promoting individual growth and, at the same time, solidifying that sense of community that is such an essential and primal human need. As Don wrote:

How, finally, can the obstacles I have detailed be surmounted and spiritual community be attained? Within communities, I believe it can be achieved only through transcendent vision, which is the sense within each person that everyone in the community is a part of everyone else—have or have-not, those initiated or less initiated, male or female, straight or gay, black or white, Jew or gentile, child, teen, younger adult, or senior citizen—and that beyond this oneness all are part of and inseparable from the rest of the animate inanimate universe. This vision is not something that can be monopolized by an elite, but a potentiality that, in Quaker terms, is "that of God in every person." Only thus can a community become a community, and its members realize their own salvation through the unfolding of their power to save the rest of the universe.

This doesn't mean that a community must be spiritual or religious to succeed. Only 35 percent of the communities listed in *The Communities Directory* are overtly religious or spiritual. Similarly, it doesn't mean that a hierarchical leadership style is necessary: only 9 percent list themselves as having a hierarchical or authoritarian structure, and the majority are democratic. But the nature of a healthy community is one that nurtures the individual and family. For many people, that, itself, is a definition of one important type of spirituality.

A life's mission or work

While a mission or lifework may seem to be the same as having a strong shared vision, it really goes a step beyond that. The most successful communities I have known over the years are those with a shared vision that is put into *action*.

This ranges from the lifetime mission of enlightenment as seen in meditation communities, to Christian lifestyle as seen in the communally organized Christian communities (which are among the fastest growing of any type of group). It also includes communities organized around a specific mission such as caring for abused children (as was the Salem Children's Village's work, a community that

Louise and I started in 1978). My observation has been that for a community, *work is important.*

The work of the community serves as a galvanizing point, a shared effort that is, in the simple act of doing it, a living-out of the vision. It's a daily reminder, a daily step toward the (usually unattainable in a single lifetime) goal as seen in the community's vision.

There are parallels to this in individual life as well. When a person doesn't feel a sense of mission or purpose in their work, their life often slides, rudderless, toward a dull cardboard-like existence. They escape into drugs, television, alcohol, or behaviors like compulsive sexuality or gambling. They're essentially lost. At the end of their lives, they look back with deep regret.

On the other hand, people who have a sense of mission about their work are happier, more motivated, more productive, and more likely to remain healthy, both physically, emotionally, psychologically, and spiritually. This link between work and a sense of mission or purpose has been documented in dozens—perhaps hundreds—of psychological studies and industrial analyses in the workplace over the past 50 years. It shows one of the cardinal points about how to create and maintain successful communities.

Shared survival concerns

The world and its inhabitants are facing some very serious problems. The beginning of the twenty-first century is not, of course, the first time this has been the case. Since the beginning of humankind—and certainly in times of difficulty or tribulation—people have banded together into community to ensure their and their families' survival.

The history of intentional communities in the United States demonstrates this: there were peaks of community-building activity during the depression that followed the Civil War, during the Great Depression, and during the 1960s when many young people thought their government had gone mad as it sent over 50,000 of them to their deaths in Vietnam. It seems that another wave of community-building is arising now, most likely out of the shared perception that things may get difficult and banding together is a better way than to go it alone.

These types of communities, while the most common to pop up during bad times, are also often the first to die out. Once the crisis is over, their shared mission is lost, they drift, and eventually disintegrate. If they do survive, they often turn into small towns and eventually lose their sense of community. Louise and I visited one of the more famous of the "communes," which sprang up in the 1960s, and found that the land and businesses had recently been privatized (shared ownership had been one of their hallmark community values), the communal commitment to vegetarian diet was being ignored by some, and many people spent most of their time in their homes watching television (previously *verboten*). Louise and I showed up at the outdoor amphitheater famous for its Sunday morning services, and discovered we were the only people there that day: almost all the activities of the community except those having to do with governance and the annual community party had disappeared. In this respect, this once-thriving commune has essentially turned itself into a subdivision for aging hippies. Once the shared vision began to crumble (as a result of a crisis in leadership), individual concerns outweighed any sense of mission or purpose and the community turned into a small town. (You could say that they are "successful" in that they're now a "shared survival" community. Our observation of it, though, was that it is now very fragile because it's lost its leadership and its sense of mission.)

This is the danger of simply jumping into a community without going through the basic exercise of determining your own personal life goals, mission, and purpose. It's the danger faced by the founders of a community if they fail to go through a similar group exercise. When the biggest mission of the members of a community is to provide a roof over their heads, they might as well be living in suburbia. On the other hand, a community can be world-transforming when a shared vision is strong, a group mission is acted out daily as part of the life and work of the community, and people come together with shared values and purpose.

* * *

Consider how different the quality of your life might be if you and a dozen other like-minded families were to get together, pool your resources, and buy enough rural acreage that you could grow much of your own food and gather enough wood or solar energy to help heat your homes. Imagine that in the midst of this, you work together daily in some way to fulfill a mission: it might be as small as a morning prayer together that radiates out power and transformation to the planet, or as large as publishing guidebooks to help others find their way, or running a training center about healing or survival, or any of hundreds of other important topics.

Of course, this isn't limited to rural environments. The cohousing movement is strong in Europe and growing in the United States, where in cities the size of Berlin and New York groups of people are buying up or renting entire blocks of city space to live in proximity to each other, often also developing elaborate rooftop gardens or parks. In smaller and mid-sized cities it works even easier, as the success of the Ten Stones community near Burlington, Vermont, demonstrates.

The primary impediment to community is the stranglehold that hierarchical corporate-driven models of employment have on individuals. They so completely drain the life-energy of people in endeavors that are only marginally related to survival or fun that when people return home at the end of the day they're exhausted, suitable only for a few hours of TV and dropping into bed.

A primary key to successful, long-term community is that the group of people are interdependent for their survival or livelihood. It may be that they're all living and working together (like in the carnival I worked at for two weeks when I was thirteen, or the community that Louise and I started); or working together but living in separate homes/neighborhoods (as with many small businesses and, particularly, co-ops); or working in ways that provide goods and services to each other (as in small-town America a hundred-plus years ago).

Living in close proximity itself isn't enough. When this is all that happens, you've shifted from "community/tribe/town" to "subdivision," regardless of how "like-minded" are the people involved. Community doesn't require the rural land mentioned earlier. It can as easily be done with no land (as with Gypsies around the world), with a homeland (the story of the Jews for thousands of years), or even in a city or suburb with the living space up to each individual (as in co-ops or other egalitarian business ventures that provide a living for their members/owners).

Even living on the same continent isn't required for community to function. For example, I operate several forums for the Internet that each provide a meeting place and community for tens of thousands of people around the world. The "place" of these communities is virtual; the ADD Forum and the New Age Living Forum, for example, exist only as electronic impulses on the Internet, yet we have members on six continents. We know each other intimately, share our joys and failures, mourn our members who are lost to misfortune or disease, and celebrate those who experience successes in their lives. We've had several people marry after meeting on-line, and one of our closest friends and fellow Sysops (systems operators or moderators), J. B. Whitwell, recently died of lung cancer, causing a spasm of grief in our cyber-world.

So the key to community is interdependence—economic, living needs, or emotional support/friendship—not proximity. When proximity is added to the mix, you have what looks like traditional tribes, but what makes a tribe truly a tribe is not its racial identity or location but the interdependence of its members on each other. This can happen anywhere, in any form.

Many people are already taking these steps today. I see them as one of the best hopes for the survival and enlightenment of humanity as we move into times of distress and shortage. When properly structured, such communities meet vital social and spiritual needs as well as providing for the life-support needs (food, shelter, and sometimes employment) of their members.

Reinventing Our Daily Life and Rituals

 A culture without its storytellers will eventually cease to be a culture.

—ARI MA'AYAN, Muskogee Creek Native American and spirit coach, 1998

One of the most important of the "secrets" we've lost from ancient people, from the Older Cultures, is that of how to conduct ritual and ceremony. In a way, it's not altogether accurate to say we've "lost" them; we've instead turned our rituals from things that remind us of the sacredness of all life into celebrations of consumerism as if it were itself a religion. As a culture, we've removed the sacred from both daily life and what were once holy days and are now merely holidays.

It's small wonder Thoreau would write, "Most men live lives of quiet desperation." Most people do, in large part because life—this extraordinary gift we've been given, this marvelous opportunity to experience the world—has been stripped of its awe and wonder by our culture. Corporations have taken it a step further; they've built into the vast majority of the 20,000-plus TV commercials the average American sees every year the message that "the goal of life is to consume"—more, better, newer, improved, higher status. In this regard, you could say that we haven't just made our daily lives ordinary, but we've profaned them; most people now every day lay their hopes, fears, and dreams before the altar of consumerism. They hope for the better car, fear losing their job, and dream of a larger and more impressive house.

Rituals don't go away, they merely change

We do have lives filled with ritual, whether we realize it or not. There's the ritual of the morning cup of coffee, watching the morning news, reading the paper, driving to work, taking the lunch hour, watching the favorite TV shows, going to bed at night. For most people these are our "unconscious" rituals. We do them without thinking about them, without realizing that they've come to occupy so much of our lives that they've *become* our lives.

Ancient people, and the tribal people living today most often use these ordinary rituals of daily life as reminders to wake up to the extraordinary spiritual vitality of life. We have a vestige of this in the ritual of saying grace or prayers before meals, although this has largely disappeared in modern society. As we were preparing for a particular ceremony, an Apache elder told me that, as we gathered the sticks from the forest, we must ask each stick if it would like to become part of the fire to heat the stones for the sweat lodge. We must listen to the answer, and consider it seriously. Those that said "No" we must leave behind. And this wasn't something that was unique to gathering firewood for a sweat ceremony; every act, throughout each moment of the day, he said, was an opportunity to communicate with the Great Spirit who was present in and spoke through all creation.

Two weeks before writing these words, I was giving a talk and book signing at a Barnes & Noble store in a northeastern state. I spoke about attention deficit disorder (the topic of nine of my books), and the suggestions and ideas I have for how parents and children can deal with ADD and other "learning disabilities." After my talk, as I sat at a small table signing books, a couple in their early forties approached me. He was wearing a business suit; she wore fashionable dress. They explained that they'd come to the bookstore from a meeting with a client of the husband's company. Then, mostly to me but also to the dozen or so other parents who stood around the table, the woman said something that brought a sharp pain to my chest.

"You know how most people have a vacation home, a boat, take a cruise every year?" she said. I nodded, not knowing where she was going with the thought. Her voice had an angry edge to it, a bitterness like that of a person who feels life has dealt with her unfairly. "Well, Billy is our boat, our cruise, our summer home," she said. "He has ADD, and couldn't make it in school, and so we shipped him off to a boarding school, where he ran away and then got caught by the police. In the detention center he got in a fight with a guard he said had tried to rape him, and so now they've charged him with assault, and he's only sixteen and already he's a juvenile delinquent. We paid a lawyer three thousand dollars to get the court to suspend the charges, and now he's in a psychiatric hospital for children. This is all costing us—even after insurance—more than that boat or summer home would, and I don't see any end in sight until he's eighteen and out on his own. By then we could be broke."

There was a genuine anger and pain in her voice, and I could also see it in her husband's eyes as he nodded in agreement. Their child, they believed, had robbed them of the blessings promised by their society. He had stolen from them their opportunity for happiness. He'd taken away the goals of their lives that they had worked very hard for so many years to reach.

In their minds, the most sacred and meaningful rituals of life were to visit the vacation home, take the boat out on the lake, and go on an annual cruise. All I could do was look at them and say, "I hope your son makes it through this experience intact. Psychiatric hospitals can be very painful places for children."

"Oh, *he'll* make it through," the father said. "He's got his whole life ahead of him. I don't know if we'll make it through, though."

I looked around the circle of 12 or 13 parents who were waiting to have books signed, and about half of them were nodding in agreement. They, too, were worried that their children's "disorder" might steal the happiness they had counted on.

Four days earlier, I was in Germany with my mentor and friend Gottfried Müller. We walked along his favorite path through the woods—what he calls his Prophet's Way—and he pointed out a tree that was growing up from the side of the mountain. Its roots curved over and around an outcrop of rocks, coarse brown against the dark gray of the granite. "Look how the roots have found their way," he said. "There is a life, an intelligence there, a spirit."

He touched the tree with his hand, and said, "Thank you for your life, and for being here and making my life better."

I touched the tree, too, and said a silent prayer.

A few feet ahead on the trail, he stopped and pointed to the leafless early-winter trees covering the mountain across the valley. A few hundred feet below us the Steinach river danced and gurgled. The air smelled of snow, and the cold tickled my nose. "We can see this, Thomas," he said. "You can hear the river, feel the cold of the air." Tears came to his eyes. At 84, he'd often been talking about soon "passing on."

"Life is such a precious and rare gift," he said. "Think of all the people who have been alive and today are not. What would they give to stand here and breathe this air? Life is such a gift to us from God."

For Gottfried Müller, as for many Older Culture peoples, the ritual of walking through everyday life is filled with reminders of the sacredness of life. In those moments, we receive—in full measure—the happiness and meaningfulness of life that our Younger Culture consumer/corporate religion promises a thousand times a day but can never truly deliver.

Intentional rituals

In addition to the rituals of daily life—the "ordinary" rituals—we also have conscious and intentional rituals. These include such things as marriage, attending church or synagogue, and rites of passage through life.

Almost every human society has these ceremonies marking the milestones of life. In modern society, however, again these have been largely lost or transformed into "consumer festivals" run by corporate priests such as wedding consultants or department store "Christmas gift" areas and photo-op Santas.

This is particularly devastating for our children. High school graduation and joining the military are about the only pre-marriage ones left, for most children, and even these are increasingly "no big deal." Those raised in religious families have the Bar or Bat Mitzvah and Confirmation, but these are among our few "leftovers." In the headlong rush toward a "better life," they're much more the exception than the rule.

Yet our children need rituals of passage: according to the U.S. Center for Health Statistics, suicide among children in the age group of 15 to 19 *doubled* between 1970 and 1990. Suicide and homicide account for fully a third of all deaths among American teenagers.

Rituals are, in many ways, an important part of the glue that binds together a culture, society, family, or relationship. In modern society, we're witnessing a rapidly accelerating disintegration of our rituals, shifting the extra-ordinary into the ordinary. Some institutional leaders seem uncomfortable with ritual, and so feel the need to "break it up"; others are just in a hurry to get home to watch their favorite show. A friend sent me an e-mail after reading the first draft of this chapter, in which he noted:

> I guess it's not ironic that just tonight I was thinking about how rituals (even secular ones) just ain't the same anymore. Tonight my daughter was inducted into an important organization at her school. When you and I were growing up, the speaker would have discussed the values embodied by the kids who were being honored, but this speaker spent the first one-fourth of his remarks poking fun at the kids about teenagers being "not all there." I know that's not the kind of ritual you're talking about, but I think there's something in common—there's a general decline in our recognition of what's important.

In my book *The Prophet's Way*, I detailed witnessing part of a puberty ritual for an Apache girl making the transition into womanhood.[1] It is a ritual that has survived thousands of years, and clearly transformed this young girl. Similar rituals are part of every Older Culture society I've studied.

I've participated in drumming ceremonies, talking circles, sweat-lodge ceremonies, and other Native American, Aboriginal Australian, and Native African rituals on four continents; each was rich with significance both to the individuals and to the culture. One of the most common Native American rituals is the talking circle, where a sacred object (a stick, an eagle feather, or something else) is passed from person to person. Only the person holding the object can speak, and in most Native American tribes the other members of the circle respectfully avoid eye contact with the speaker, the better to focus on hearing the person's words. The object goes around the circle, and nobody interrupts or even responds to the comments or questions of others until their turn comes. If a person strongly agrees with the speaker, they may softly grunt or make a sympathetic sound in their throat, but that's the limit. When it's a person's turn to talk, usually they begin by thanking: thanking the Creator for being there, thanking their parents for their life, thanking those in the circle for listening and sharing. Only then do they begin to speak about whatever's on their mind. It's an extraordinary and bonding experience, one that trains patience as well as respect.

Ancient people understood the importance of ritual, and filled their lives with it. It was both a way of bringing predictability to an insecure world, and also of constantly reminding themselves of the presence and sacred nature of the divinity filling the world.

Dave deBronkart, one of my editors on this book, noted that in his Episcopal confirmation class he was taught, "a sacrament is an outward and visible sign of an inward and spiritual grace." Rituals aren't only for show; they *can* internalize experience. Dave added, "Rituals and ceremonies are the shared experience that becomes the vibration that we all have in common today, and that resonates with our ancestors down through the generations. I suppose it's no coincidence that when an oppressive government wants to stamp out a culture, they cut off its ceremonies . . . and if the ceremonies can still go on in secret, the culture can re-emerge when the oppressors are gone."

Reinventing rituals

If you bring rituals and ceremonies back into your life, you'll notice an immediate change in the quality of your life. The ceremonies may change from time to time, they may be wholly new inventions of your own, or they may be borrowed from ancient peoples.

For example, on most days my wife and I sit and meditate for a few minutes in the morning. We often take walks to an altar that Herr Müller built in the forest near our home, and pray there. We often say grace before meals, and a special prayer before drinking a glass of red wine with Friday's dinner, reminding us of Jesus' blood and the sacrifices He made for us. We conduct a version of the Friday-night Sabbath ceremony that Herr Müller taught us (and which his mentor taught him), and take Saturdays off whenever possible, using the day for relaxation, reading, discussion, and long walks in the forest around us. We give each other special attention each morning the first 15 minutes or so after waking up but before getting out of bed, hugging and talking about the day, and reaffirming our love for each other. All of these are rituals that we have created with intent, and through which we find the presence of the sacred.

With our friends, we are experimenting with talking circles and other ceremonies. We're considering inviting people to our home for weekly meditation periods (if my travel schedule ever slows down), as we did years ago in Michigan and New Hampshire, and building more sacred ritual into our gardening and herb-gathering.

You can create your own rituals and ceremonies, too. After a few weeks of intentional effort, they become easy and natural parts of life, yet the reminders remain.

When you live in community, the community can build a superstructure to make the rituals even easier to remember to follow, and in some ways even more meaningful. I've spent several Sabbaths and shared a Passover seder in a Hasidic community in Jerusalem: the level of detail in the rituals was both extraordinary and powerful. In the first years we ran the New England Salem Children's Village (before a social worker objected that we were being "too religious" and the state required us to stop), Grandpa Irving would ring a bell every morning around nine and every afternoon around three. Those adults (and the occasional child) who chose to join us would gather in my office for 15 minutes of silent prayer and meditation, which we ended by holding hands in a circle and saying the Lord's Prayer. Those communities organized around specific goals—to heal others and/or to save the planet, for example, and certainly those devoted to religious practice—find many ways to build ritual into their lives.

In the saying of thanks for a meal or the daily walk in the forest—or even down the suburban street, noticing and acknowledging the grass and bushes and trees—you reconnect with your ancient ancestors. By doing this, you bring forward into the present and the future their wisdom, their sustainable way of life, their view of the world, which you can then share with others. In these, we may find our greatest opportunity to transform our Younger Culture.

Transforming Culture Through Politics

 The care of human life and happiness and not their destruction is the first and only legitimate object of good government.
—THOMAS JEFFERSON (1743–1862), 1809

While much of this book has been devoted to understanding and defining the differences in worldviews and behaviors between Older and Younger Cultures in the largest or most macro sense, politics is an arena where a clash of these worldviews becomes so close and real, it impacts our daily lives. As such, it's worth examining in some detail.

Younger Culture politics has always been the politics of domination, war, and conquest. Dominator leaders claim that they are the holders of rights, and all under them have only privileges. The leaders variously claim they come by these rights to rule through their willingness to use violence, through the blessings of gods they alone can speak with or on behalf of, or because their great wealth is a sign of either moral rectitude or divine selection.

Leaders and political institutions of this kind constitute more than half of the governments of the world, and are described as kingdoms, dictatorships, monarchies, single-party-rule states, and the many variations of corrupted states that are called sham democracies. A powerful elite that controls the nation's natural resources and is armed to the teeth holds together most such overtly Younger Culture nations.

A little over two thousand years ago, the citizens of Athens experimented with a different form of governance, in which the people themselves held the power of the state. To ensure no one individual, family, or group would acquire unending power or government-funded wealth, the politicians of Athens were selected annually by lottery, the way jury duty is done today. When your name was drawn, you had to serve for a year, and then you were out again. With no voting, no campaigning, and little corruption, the Athenian system became progressively more and more democratic over its two-hundred-plus years of existence, and eventually even many of the island's slaves were enfranchised.

Unfortunately, the Athenian experiment with Older Culture values came to a sudden halt in 338 B.C.E., when Alexander the Great conquered the island. Over the next two thousand years, small experiments in Older Culture values appeared briefly in Florence and Venice, and apparently flowered among the Saxons in England during the roughly six hundred years between the time when the Roman Empire fell and the time they were again conquered by the French from Normandy at the Battle of Hastings in 1066.

So when the Founders of America set out to try to create a nation-state using Older Culture values, there weren't many good role models in the history of what was then called the civilized world. This led many of the Founders, the most well-known being Franklin and Jefferson, to examine closely the forms of governance used by the Native American tribes of North America, and the systems of governance used prior to the Norman invasion in ancient England. Synthesizing their histories and observations of these with the revolutionary new writings of people like John Locke, they came up with the firm conviction that it was possible for "civilized" city-state people to form a government that held largely to Older Culture values.

As I've already described, the primary value was that the purpose of community and government is to provide for the "life, liberty, and pursuit of happiness" of its people. Second among them was that only individual citizens were the holders of rights. They rejected the historic systems that claimed rights derived from heredity (kings) or divine blessing (theocracies) or wealth (feudalism), and said instead that *all* forms of human association—be they businesses, churches, or even governments themselves—would have only privileges, and those privileges would be defined by "We, The People," the sole holders of the rights.

Thus was born modern democracy, and within a decade the idea had crossed the Atlantic and infected the French, and soon other nations as well. From 1776 to the 1980s, democracy slowly but steadily spread across the globe, in most cases and places carrying with it many powerful embedded Older Culture notions.

However, even from the beginning there were doubters, those who held to Younger Culture notions of "might makes right" and that original sinner humans must be under the tight control of government if they were to behave correctly. This paternalistic worldview sharply contrasted with the Jeffersonian egalitarian worldview, and in the 1790s these contrasts caused Jefferson and his colleagues to split from Federalists Adams and Hamilton—who were steadily moving more and more in a Younger Culture direction—and form the Democratic Republican Party (now known as the Democrats).

The Federalists, open advocates of a powerful central government dominated by a ruling elite, so fought among themselves that in the early 1800s they melted down, and were replaced by the Whigs, who shared a similar perspective. In the decades prior to the Civil War, the Whigs also disintegrated from internal power struggles, and were replaced by the Republicans, who, in the years following Lincoln's assassination, came out openly as the party that represented the interests of the wealthy and the corporations.

Thus, in many ways, the split between the Republicans and Democrats could be seen more as a clash of worldviews than of politics. The politics make concrete the specifics of the different worldviews. In the past few decades, as the Democrats have also moved closer to the supporting corporate interests over human interests and the Supreme Court has empowered corporations to corrupt politics with campaign cash, other parties have emerged (Greens, Progressives) to fill the relative void in Older Culture political perspectives.

The Younger Culture "new" conservatives

As we passed into the twenty-first century, democracy around the world as a preferred form of government has become increasingly imperiled by corporate rule. Those advocating the use of preemptive force and an open and unashamed rule by the rich have come to be known as the new conservatives, or neoconservatives. They reflect an almost pure Younger Culture worldview, one tinged with fear and driven by the need to dominate and control. In the wake of the sudden neoconservative rise to power with the Supreme Court's selection of George W. Bush as president in 2000, many Americans wonder why the neoconservatives seem so intent on crippling local, state, and federal governments by starving them of funds and creating huge federal debt that our children will have to repay.

Many think it's just to fund tax cuts and subsidies for the rich, that the multimillionaire CEOs who've taken over almost all senior posts in government are just pigs at the trough, and this is a spectacular but ordinary form of self-serving corruption. It all seems so plausible, and there's even a grain of truth to it. But juicy deals for right-wing government insiders and their friends are just a byproduct of the real and deeper war against democracy. The neoconservatives are perfectly happy for us to think they're just opportunists skirting the edges of legality and morality, but this is far more dangerous than simple government corruption.

Indeed, the neoconservatives claim to be anti-government. As a leading spokesman for the neocon agenda, Grover Norquist, told National Public Radio's Mara Liasson in a May 25, 2001, *Morning Edition* interview, "I don't want to abolish government. I simply want to reduce it to the size where I can drag it into the bathroom and drown it in the bathtub."

Without a larger view, the issues of domestic spending, oil, neoconservative power plays in both major parties, the loss of liberties, anti-government rhetoric, and war in the Middle East all seem like separate and unconnected events. They're not. The "new conservatives" who've seized the Republican Party, and who, through the Democratic Leadership Council (DLC) are nipping at the heels of the Democratic Party, are not our parents' conservatives. Historic conservatives like Barry Goldwater, Harry Truman, and Dwight Eisenhower would be appalled. Although their philosophical roots go back to Alexander Hamilton, who openly argued during the Constitutional Convention that royalty was the best form of government, the neocons have always been kept to the fringe.

Indeed, the "Reagan/Bush Revolution" flew in the face of traditional conservative ideals. As John Stockwell notes in his book *The Praetorian Guard: The U.S. Role in the New World Order*, Reagan/Bush were proud of their contempt for the concerns of environmentalists, with Reagan once saying, "If you've seen one redwood, you've seen them all."[1] Their Department of the Interior under James Watt sold off minerals and forests to campaign contributors at fire-sale prices, and their EPA, in many cases, moved from prosecuting corporate polluters to legitimizing and protecting them under the guise of "regulation."

Although James Madison wrote in 1792 that an important role of government was to promote a strong middle class "by the silent operation of the laws, which, without violating the rights of property, reduce extreme wealth toward a state of mediocrity, and raise extreme indigence toward a state of comfort," that wasn't a sentiment shared by those in the Reagan/Bush Revolution. Instead, Reagan raised taxes on the middle class and working poor while cutting taxes by more than 60 percent for the most wealthy in America. At the same time, he bragged that he'd eliminated more than a thousand programs for poor people, and even proposed that poor schoolchildren should be content with ketchup as their daily vegetable.

At the same time, the Reagan/Bush administration, and later the George W. Bush administration, worked hard to roll back the very individual liberties that America's Founders had fought and died for. Dwight Eisenhower left office warning Americans about the dangers of the concentration of power resulting from corporations getting into bed with the military, but the Reagan/Bush and Bush administrations openly embraced these corporate powers, inviting them into the halls of governance and hungrily sucking at the teat of their campaign contributions.

In the past those promoting what is now called the new conservative agenda went by different names.

The Founders of America knew that for six thousand years "civilized" humans had been ruled by one of three groups: kings, theocrats, or feudal lords. Kings held power by virtue of the threat of violence and continual warfare; theocrats and popes held power by the people's fear of a god or gods; and feudal lords by wealth and the power that comes from throwing average people into poverty.

The "new" idea of our Founders in 1776 was to throw off all three of these historic tyrannies and replace them with a fourth way—people being ruled by themselves, a government that derived its legitimacy and continuing existence solely from the approval of its citizens. Government of, by, and for "We, The People." They called it a constitutional republican democracy.

The new feudalism

What we are seeing now in the conservative agenda is nothing less than an attempt to overthrow republican democracy and replace it with a worldwide feudal state.

The last time this happened, the feudalists took over a monarchy and then North America. In December 1600, Queen Elizabeth I chartered the East India Company, ultimately leading to a corporate takeover of the Americans that the colonists ended with the Boston Tea Party and, three years later, the American Revolution. The corporate-state partnership of the East India Company and the U.K. went on to conquer India, but eventually disintegrated as the British Empire faded, and the British government, along with most of western Europe, embraced somewhat more Jeffersonian forms of democracy.

Conservatism raised its head again in the twentieth century, revived by Franco, Hitler, and Mussolini. The Italian dictator even used the word "corporatism" to describe it, and then later renamed it as "fascism"—a word that was defined in American dictionaries such as *The American Heritage Dictionary* (Houghton Mifflin Company) in 1983 as "A system of government that exercises a dictatorship of the extreme right, typically through the merging of state and business leadership, together with belligerent nationalism."

Since the Reagan/Bush Revolution, two centuries after we rose up and rebelled against the British, this ancient enemy of democracy is again trying to seize America. The Reagan/Bush administration ignored the Sherman Act and other restraints on corporations, and sold at fire-sale prices the airwaves once held in common by "We, The People." The result was predictable: a merger and acquisitions frenzy, and the takeover of American media by a handful of mega-corporations. Bill Clinton then helped export corporatism to the industrialized world when he pushed through Congress NAFTA and GATT/WTO—treaties that raised corporations to a level of power equal to and in some cases superior to sovereign governments.

Thus, the war on Iraq was just one front in the larger feudal war against democracy itself. (And a particularly useful one—it gave the corporate feudal lords access to oil wealth, and was as effective as planned at distracting the populace from Bush's outrageous domestic agenda.)

In 1936—years before America turned its attention to fighting fascism in Germany—Franklin D. Roosevelt was concerned about the rise of a corporate feudalism here in the United States. In a speech in Philadelphia on June 27, he said: "Out of this modern civilization economic royalists carved new dynasties. New kingdoms were built upon concentration of control over material things. Through new uses of corporations, banks and securities, new machinery of industry and agriculture, of labor and capital—all undreamed of by the Fathers—the whole structure of modern life was impressed into this royal service."

Roosevelt suggested that human nature may play a part in it all, but that didn't make it tolerable. "It was natural and perhaps human," he said, "that the privileged princes of these new economic dynasties, thirsting for power, reached out for control over government itself." It was a control the Democratic Party of 1936 found intolerable. "As a result," Roosevelt said, "the average man once more confronts the problem that faced the Minute Man."

Republicans of the day lashed out in the press and on radio, charging that Roosevelt was anti-American, even Communist. Without a moment's hesitation, he threw it back in their faces. "These economic royalists complain that we seek to overthrow the institutions of America," Roosevelt thundered in that 1936 speech. "What they really complain of is that we seek to take away their power. Our allegiance to American institutions requires the overthrow of this kind of power. In vain they seek to hide behind the flag and the Constitution. In their blindness they forget what the flag and the Constitution stand for."

The sacred archetype

Those of us who still believe in republican democracy would have "We, The People" make the decisions through representatives we've elected without the feudal influence of corporate money. We realize that "big government" is, indeed, a menace when it's no longer responsive to its own people, as happened in Germany and Russia in the last century—and is happening today in America under the neoconservatives.

But we also remember the vision of a free and democratic America—a sacred archetype so powerful that protestors in Tiananmen Square marched to their deaths carrying a 36-foot-tall papier-mâché replica of the Statue of Liberty while quoting the words of Thomas Jefferson.

Facing the power of the East India Company's corporate feudalism in 1773, the Founders of our nation, unable to get their voices heard in the halls of the British government or even in many of the newspapers of the day, turned to two nonviolent and very effective methods to spread the new meme of democracy.

The first was pamphleteering—and the Internet is today's pamphlet. Millions are using e-mail and pointing to websites to awaken people and promote democratic change.

The second was creating "committees of correspondence," also used extensively by the Women's Suffrage movement. These were groups organized to write letters to the editors of newspapers.

People across America have already begun letter-writing, faxing, and e-mail campaigns, and you can see the results on the editorial pages of our newspapers and in the reactions of some of our politicians. Other correspondents are blogging on the Internet or calling in to radio talk shows, modern variations on this theme.

A correspondent in York, New York, who is pamphleteering in e-mail and encouraging committees of correspondence to write letters to newspaper editors against the new feudalism's wars on America and overseas, shared the following quote from Emerson: "One of the illusions [of life] is that the present hour is not the critical, decisive hour."

Yet this is the critical and decisive hour, and we are not without voices or tools.

Taking Back America

Only those who dare to fail greatly can ever achieve greatly. The future does not belong to those who are content with today, apathetic toward common problems and their fellow man alike, timid and fearful in the face of bold projects and new ideas. Rather, it will belong to those who can blend passion, reason and courage in a personal commitment to the ideals of American society.
—ROBERT F. KENNEDY (1925–1968), 1966

Some consider partisan politics to be "just part of the game," problem rather than solution, or beneath the lofty gaze of those concerned with worldwide ecocide and both species-wide and human genocides. The Founders of the United States, however, thought of government as a sacred vehicle and embraced partisan politics as a way of furthering that view.

I believe it's not just important but vital that Americans awaken to the crises in our body politic and begin the urgent work of restoring political power to "We, The People." This is, after all, the most essential of the ancient ways humans have lived—in tribal egalitarian democratic forms.

Thus, we face this modern crisis, which contains within it the seeds of great opportunity. Marching in the streets and speaking out for the sacred and ancient human form of governance known as democracy is important work, but wouldn't we have greater success if we also took control of the United States government? It's vital to point out right-wing-slanted reporting in the corporate media, but isn't it also important to seize enough political power in Washington to enforce anti-trust laws to break up media monopolies?

And how are those of us carrying an Older Culture vision—most standing on the outside of government, looking in—to deal with oil wars, endemic corporate cronyism, slashed environmental regulations, corporate-controlled voting machines, the devastation of America's natural areas, the fouling of our air and waters, and an administration that daily gives the pharmaceutical, HMO, banking, and insurance industries whatever they want regardless of how many people are harmed?

This lack of political power is a crisis others have faced before. Americans who care about the future of democracy and believe there is a deep and real spiritual component to politics should learn from their experience.

After the crushing defeat of Barry Goldwater in 1964, a similar crisis faced a loose coalition of gun lovers, abortion foes, southern segregationists, Ayn Rand libertarians, proto-Moonies, rich neoconservatives, and those who feared immigration within and communism without would destroy the America they loved. Each of these various groups had tried their own "direct action" tactics, from demonstrations to pamphleteering to organizing to fielding candidates. None had succeeded in gaining mainstream recognition or affecting American political processes. If anything, their efforts instead had led to their being branded as special-interest or fringe groups, which further diminished their political power.

Don't get angry. . . .

As a result of these defeats, the neoconservatives decided not to get angry, but to get power. Led by Joseph Coors and a handful of other ultra-rich funders, and funded by literally millions of $10 contributions collected by the likes of Jerry Falwell, they decided the only way to seize control of the American political agenda was to infiltrate and take over one of the two national political parties, using their own think tanks like the Coors-funded Heritage Foundation to mold public opinion along the way. Now they regularly get their spokespeople on radio and television talk shows and newscasts, and write a steady stream of daily op-ed pieces for national newspapers. They launched an aggressive takeover of Dwight Eisenhower's "moderate" Republican Party, opening up the "big tent" to invite in groups that had previously been considered on the fringe. Archconservative neo-Christians who argue the Bible should replace the Constitution even funded the startup of a corporation to manufacture computer-controlled voting machines, which are now installed across the nation. And former arms manufacturer, convict, and multimillionaire businessman Reverend Moon took over the *Washington Times* newspaper and UPI.

Their efforts have borne fruit, as Kevin Phillips predicted they would in his prescient 1969 book, *The Emerging Republican Majority*, and as David Brock so well documents in his book *Blinded by the Right*.[1]

But the sweet victory of the neoconservatives in capturing control of the Republican Party, and thus of American politics, has turned bitter in the mouths of the average American and humans around the world. Thus, many are suggesting that it's time for concerned Americans to reclaim Thomas Jefferson's Democratic Party. It may, in fact, be our only short-term hope to avoid a final total fascistic takeover of America and a third world war.

"But wait!" say the Independents, Greens, Progressives, and left-leaning Reform Party members. "The Democrats have just become weaker versions of the Republicans!" True enough, in many cases. And it isn't working for them, because, as Democrat Harry Truman said, "When voters are given a choice between voting for a Republican, or a Democrat who acts like a Republican, they'll vote for the Republican every time." (History shows that voters are equally uninterested in Republicans who act like Democrats.)

Alternative parties have an important place in American politics, and those in them should continue to work for their strength and vitality. They're essential as incubators of ideas and nexus points for activism. Those on the right learned this lesson well, as many groups that at times in the past had fielded their own candidates remain intact and have also become powerful influencers of the Republican

Party. Similarly, being a Green doesn't mean you can't also be a Democrat.

But what about ideological "purity"?

That America was designed as a two-party state is not a popular truth.

There's a long list of people who didn't like it—Teddy Roosevelt, H. Ross Perot, John Anderson, Pat Buchanan, Ralph Nader—but nonetheless the American Constitution was written in a way that allows for only two political parties. Whenever a third party emerges, it's guaranteed to harm the party most closely aligned to it. This was the result of a well-intentioned accident that most Americans fail to understand when looking at the thriving third, fourth, and fifth parties of democracies such as Germany, India, or Israel. How do they do it? And why can't we have third parties here?

The reason is because in America we have regional "winner take all" types of elections, rather than proportional representation where the group with, say, 30 percent of the vote, would end up with 30 percent of the seats in government. It's a critical flaw built into our system, so well identified in Robert A. Dahl's brilliant book *How Democratic Is the American Constitution?*[2]

When the delegates assembled in Philadelphia in 1787 to craft a constitution, republican democracy had never before been tried anywhere in what was known as the "civilized world." There were also, at that moment, no political parties, and "father of the Constitution" James Madison warned loudly in Federalist #10 against their ever emerging. In part, Madison issued his warning because he knew that the system they were creating would, in the presence of political parties, rapidly become far less democratic. In the regional winner-take-all type of elections the Framers wrote into the Constitution, the loser in a two-party race—even if s/he had fully 49.9 percent of the vote—would end up with no voice whatsoever. And the combined losers in a three-or-more-party race could even be the candidates or parties whose overall position was most closely embraced by the majority of the people.

The best solution to this unfairness, in 1787, was to speak out against the formation of political parties ("factions"), as Madison did at length and in several venues. But within a decade of the Constitution's ratification, Jefferson's split with Adams had led to the emergence of two strong political parties, and the problems Madison foresaw began and are with us to this day.

This is particularly problematic in presidential elections. H. Ross Perot's participation in the 1992 election drew enough votes away from the elder George Bush that Bill Clinton won without a true majority. Similarly, Ralph Nader's participation in the 2000 election drew enough votes away from Al Gore that it was easy for the Supreme Court and Jeb Bush to deflect media notice away from Florida's illegal vote-rigging in the pre-election purging of the voter rolls and thus select George W. Bush as president.

Conservative activists recognized this inherent flaw in the electoral system of the United States and decided to do something about it, recruiting Ronald Reagan and forming his infamous "kitchen cabinet." They took over the Republican Party and then successfully seized control of the government of the United States of America. As we can see by comparing documents from the 1990s *Project for a New American Century* with today's war in Iraq, these once-marginalized conservative ideologues are the real power behind Bush's throne.

Social liberals who embrace Older Culture values weren't so practically-minded. Instead of funding think tanks to influence public opinion, subsidizing radio and TV talk show hosts nationwide, and working to take over the Democratic Party, many left to create their own parties while most gave up on mainstream politics altogether. The remaining Democrats were caught in the awkward position of having to try to embrace the same corporate donors as the Republicans, although they weren't anywhere near as successful as Republicans because they hadn't (and haven't) so fully sold out to corporate and wealthy interests.

We see the result in races across the nation, such as my state of Vermont. In the 2002 election for governor and lieutenant governor, the people who voted for the Democratic and Progressive candidates constituted a clear majority. Nonetheless, the Republican candidates became governor and lieutenant governor with 45 percent and 41 percent of the vote respectively because each had more votes than his Democratic or Progressive opponents alone. (Example: Republican Brian Dubie, 41 percent, Democrat Peter Shumlin, 32 percent; Progressive Anthony Pollina, 25 percent. The Republican "won.")

Similarly, Republicans have overtly used third-party participation on the left to their advantage. In a July 12, 2002, story in the *Washington Post* titled "GOP Figure Behind Greens Offer, N.M. Official Says," *Post* writer Thomas B. Edsall noted that: "The chairman of the Republican Party of New Mexico said yesterday he was approached by a GOP figure who asked him to offer the state Green Party at least $100,000 to run candidates in two contested congressional districts in an effort to divide the Democratic vote." The Republicans well understand—and carefully use—the fact that in the American electoral system a third-party candidate will always harm the major-party candidate with whom s/he is most closely aligned.

The Australians solved this problem in the last decade by instituting a nationwide variation on what's known in the U.S. as instant run-off voting (IRV), a system that is making inroads in communities across the United States. There are also efforts to reform our electoral system along the lines of other democratic nations, instituting proportional representation systems such as first proposed by John Stuart Mill in 1861 and now adopted by almost every democracy in the world except the U.S., Australia, Greece, the United Kingdom, and Canada.

Political parties can bring about such transformation only if we, in massive numbers, join them, embrace them, and ultimately gain a powerful and decisive voice in their policy-making and selection of candidates.

Something Will Save Us

 But how did we do this? How could we drink up the sea? Who gave us the sponge to wipe away the entire horizon? What were we doing when we unchained this earth from its sun? Whither is it moving now? Whither are we moving? . . . Do we hear nothing as yet of the noise of the gravediggers who are burying God?

—FRIEDRICH NIETZSCHE (1849–1900),
The Gay Science (1882)

Wendy Kaminer wrote a brilliant book entitled *I'm Dysfunctional, You're Dysfunctional*. In it, she pointed to the pervasive assumptions of dysfunction inherent in the self-help movement, and the increasing obsession with emotional and psychological pathology in our culture. She didn't offer any specific solutions; she had only defined the problem. (Although one could say that her solution was really the most elegant of all: see the problem for what it is, and refuse to dance the dance. In this, she argued forcibly for people reclaiming their own inherent power and emotional health.)

Kaminer received numerous letters from people demanding solutions to the problems she had identified. She pointed out this irony in a later edition of the book: it was as if the people writing wanted her to suggest the creation of a self-help group or book to help those addicted to them.

Some of the initial responses to the early editions of this book were similar. I received letters, e-mails, and calls from people telling me with great certainty that the only solution to the problems outlined in the first third of the book would be found in smaller families, cold fusion, coaxing the flying saucer people out of hiding, a worldwide conversion to Christianity (at least a half-dozen different people suggested that only their particular Christian sect could bring this about, and all other Christians must ultimately recognize the error of their ways), Islam, some other religion, or the immediate institution of a benevolent one-world government. The letters ranged from amazement to outrage that I'd failed to see and support their perspective.

But these are all Something-Will-Save-Us solutions. This kind of thinking is a symptom of our Younger Culture—and fighting fire with fire is only rarely successful. Usually, it just produces more flames. As Jesus, Gandhi, and Martin Luther King Jr. demonstrated, often the most powerful and effective way to "fight back" against the pathological kings and kingdoms is to walk away from the kings, see the situation for what it is, and stop playing the dominator's game.

But that involves a shift of perspective that some people find very difficult. There are, for example, those who point to the foundational belief of our culture (and, particularly, to European-ancestry citizens of the United States) that we can solve any problem if we just put our minds to it. Some even argue that the exploding human population is a good thing, because the more people there are, the greater the possibility we will find among them the next Edison, Jefferson, or Einstein, who will figure out how to get us out of this mess. It is, of course, a simplistic, and ultimately cruel, notion, but one that has been used for years, usually to advance a dominator religious or economic agenda.

In fact, it's somewhere between unlikely and impossible that children born into the contemporary slums of Islamabad or Haiti, or even Baltimore or East Los Angeles, will grow up to change the world or solve our problems. They may become very competent; any corrections officer can tell you there are geniuses among our cities' gang members and in our prisons. But grinding poverty and pervasive violence—born of overcrowding and a lack of resources and security—rarely produce more than a surfeit of ingenious criminals and competent jailhouse lawyers.

On the other hand, Jefferson was a member of the land-owning elite, what we would today call the very wealthy. Translated into today's dollars, nearly every signer of the Declaration of Independence was a millionaire or multimillionaire. Einstein was never truly poor, and lived a life ranging from comfortable to wealthy. And even Edison, penniless when he ran away from home at age 15, entered a world with a total population that was a fifth of what it is today, rich with cheap natural resources and almost limitless opportunity for ambitious white young men who spoke American English. If any of them were to be born into the modern-day sewers of Bogotá, they might end up being hunted for sport—but it's unlikely that they'd ever have access to the resources necessary to create lasting and meaningful changes in the world.

True change is not a simple process

There is, of course, no shortage of do-this-and-everything-will-be-okay solutions proffered in the press and books. The more commonly touted include worldwide birth control, strong controls on corporate exploiters and polluters, five-dollar-a-gallon (or more) taxes on gasoline and oil products, doubling or tripling of the cost of water and electricity by increased taxation, worldwide destruction of weapons of war, more money for environmental remediation, and the creation and empowerment of new political parties not beholden to corporate powers. I even dance around the edge of such solutions in the chapter about using our current oil supplies to create non-oil-consuming energy sources such as solar; I also, however, make it clear that this is merely a stopgap.

Those who are concerned that this book doesn't emphasize technological or political solutions have—if I may say it gently—missed the point.

Missing the point of a book like this is quite easy to do, because the book makes a radical departure from the normal fare of self-help and environmentalism. It presents the problems, delves into the cause of them, and then presents as a solution something that many may think couldn't possibly be a solution because it seems unfathomably difficult: change our culture, beginning with yourself.

Such a solution is one of the most perplexing to grasp because culture, at its core, is invisible. Like the air we breathe and walk through, its presence is felt only when it's resisted. At all other times it's part of the everything-that-seems-like-nothing-around-us that we rarely consider and almost never question.

The idea of cultural change is often unpalatable because any sort of real, individual, personal change in beliefs and behaviors is so difficult as to be one of the rarest events we ever experience in our own lives or witness among those we know. It's easy to send $10 off to the Sierra Club; it's infinitely more difficult to reconsider beliefs and behaviors held since childhood, and then change your way of life to one based on that new understanding, new viewpoint, or new story. But if such deep change is what we really need, I see no point in pretending that something simpler will do it.

The something-will-save-us viewpoint

We are members of a culture that asserts that humans are at the top of a pyramid of creation and evolution. In our modern techno naiveté, we reveal our fatal belief that anything we have done—for better or worse—can also be undone. We tend to think that every problem, including man-made ones, has a solution.

In the deus ex machina ending in Greek plays, the hero inevitably finds himself in an impossible situation. To close the show, a platform is cranked down from the ceiling with a god on it who waves his staff and makes everything well again. Similarly, we have faith that somehow things will turn out okay. "Don't worry," our sitcom culture tells us, "human ingenuity will save us."

We envision that our salvation will come from new technologies, or perhaps the rise of a new leader or political party, or the return/appearance of ancient founders of our largest religions. The more esoteric among us suggest that people from outer space will show up and either share their planet-saving technology or take us to a less polluted and more paradisiacal planet. The Christian "rapture" envisions the world's "good people" being removed from this mess we've created and relocated to a paradise created just for them. Among the New Age movement, a popular notion is that just in the nick of time the Ancient Ones, now only available in channeled form through our mediums and psychics, will make themselves known and tell us how to solve our problems. And, of course, there is no shortage of "just follow me, worship me, do as I say, and you'll be happy forever" gurus.

Whatever form it takes, our culture whispers in our ears daily, "Something or someone will save us. Just continue your life as it was, and keep on consuming, because you couldn't possibly save the world, but somebody else will."

This is what I refer to as something-will-save-us thinking.

It's built into our culture, at the foundation of our certainty about how life should be lived, how the world works, and our role in it. It originated, most likely, as a way for dominators in emerging Younger Cultures to control their slaves: "Just keep picking that cotton and praying, and you'll eventually be saved. It may be after you die, but it'll happen, don't worry about that. But, in the meantime, don't stop picking that cotton!"

Far from being the solution, something-will-save-us thinking is at the root of our problems.

Younger Cultures and something-will-save-us beliefs

Something-will-save-us beliefs are at the core of Younger Cultures, but startlingly rare among Older Cultures. This is not to say Older Cultures don't have spirituality, belief in deities or spirits, elaborate ritual, offerings or oblations to gods or spirits, personal mystical experience, and so on. But Younger Culture beliefs require two essential elements that are lacking from most Older Cultures:

1. The belief there is, to paraphrase Daniel Quinn, only One Right Way to Live (which, of course, is "our" way), and that when everybody on the planet figures this out and lives our way, then things will be good. Conversely, this belief says that if we fail to convert everybody to our way of life, the deity (or, for secularists, the science/technology) who defined this One Right Way of Life will punish us. This punishment may be personal or it may involve the

destruction of the entire planet. But in either case, those who fail to conform to the dominator culture's way will suffer, and the only way to be saved from doom is to conform.

2.   The belief that humans are essentially flawed, sinful, damned by a specific deity, or intrinsically destructive, and, therefore, they (we) can and must be "saved." According to this belief, this personal (and, thus, worldwide) salvation process can only happen by either intense personal effort and devotion to a particular program (yoga, rosary, prostrations, good deeds, psychotherapy, jihad, Prozac, evangelism), or through the intervention of a divine being or beings who reside in a non-Earthly realm (aliens from space) or non-physical realm (gods, saviors, angels, prophets, gurus, channeled Wise Ones).

The most secular among us believe we will find, among our own human race, people who will save us from ourselves. Historically, this was the basis of the rule of dominator kings: they had to have absolute power over their people, they said, to save the people from themselves. This is also a core belief found among modern people who treat either politics or science as a something-will-save-us religion.

Because members of Older Cultures assume there are Many Right Ways to Live, each unique to a particular place, time, and people, they avoid evangelism. Instead, they respect other cultures and beliefs, carefully protecting their ways and beliefs from outsiders, and accepting "converts" only in the rarest of circumstances.

Believing in the flawed or "fallen" nature of humanity allows people to rationalize the various genocides, past and present, committed against humans and non-humans. According to this worldview, some of us will act out "human nature" (whether it's biologically caused as the neo-Darwinianists suggest, or a curse from an upset god as some religions suggest) and commit all sorts of crimes against the human and natural world.

But if evil is fundamental to human nature, how could it be that it doesn't exist in all cultures? Few ever pause to question whether the evil or dysfunction may be in the nature of our culture, rather than the humans who are in it.

If we could just find the right lever

Something-will-save-us beliefs—whether rooted in technology or religion—suggest that our problems are always solvable by new and improved human actions: they're things we can control and manipulate, if only we have the right levers/science or can figure out the right prayers to motivate the right god(s) or space aliens.

The technological something-will-save-us believers say that we haven't yet mastered the technology of efficient and non-polluting energy use, equitable economic and/or political systems, simple and widespread methods of food or birth control (or distribution of them), better medicines, or efficient communications. Their refrain always begins, "If only there were more of . . ." or, "If only everybody would . . ." and is then followed with the particular doxology of the particular solution being recommended.

Religionists say we just haven't yet mastered the technology of pleasing the particular god of their sect: if every last tribe is found and converted to a particular institutionalized religion, or if all the ancient prophecies are fulfilled, or if enough people would meditate with the right technique or say the right magic words or the right magical name, then we'll be saved from doom. We haven't yet gotten that system perfect, they feel, so we need to work harder on it.

Older Cultures and the synergist worldview

The true problem we're facing is a natural and predictable result of this way of viewing the world. The problem is the stories we tell ourselves, what we see and hear and feel as we move through the world, our disconnection from the sacred natural world, and our insistence on quick-fix/external-to-us solutions to natural-world crises that we ourselves created.

Most of us can't even imagine what it would be like to live with a different worldview from our own. (We do, though, keep getting glimpses, most often in the words of our "enlightened ones"—and we usually ignore those glimpses because, being Older Culture wisdom, they're so inconsistent with our way of life.)

The Younger Culture says, "Who cares what our children's children will inherit: that's their problem, and they can work out their own salvation just as we must work out ours."

The Older Culture perspective says, "We're here, now, and must deal with the practical realities of this life. Any decisions we make must consider their impact on our grandchildren seven or more generations from now."

I find value in many of the technological suggestions people are exploring and promoting worldwide, and many must ultimately play a role in the transformation of our world if we are to avoid utter disaster. But none attacks the problem at its core. We must begin to live a sustainable, egalitarian, peaceful way of life. This can happen through political or religious transformation, but at its core it's cultural transformation.

This is not a secret: Older Culture people have been shouting them to us since we first began our genocide against them seven thousand years ago. Most of them are still trying as hard as they can, but we're not capable of hearing, because our culture has plugged our ears to their message. Here it is:

> Return to the ancient and honest ways in which humans participated in the web of life on the Earth, seeing yourselves and all things as sacred and interpenetrated. Listen to the voice of all life, and feel the heartbeat of Mother Earth.

Living from this place, all other decisions we make will be appropriate.

The good news is that this is a very clear solution, embodying, as it does, only a single issue and a single change in a single culture (ours). The bad news is that that single issue is the most difficult and wrenching change I can envision . . . but we must begin, now, to take the first steps.

It's the same problem the prophets of old wrestled with: their message was most often, "Change your way of seeing and living in the world, because the path you're currently walking will lead to disaster." As secular and Bible history show, such prophets were almost always ignored, at least until the predicted (and inevitable) disasters struck, and even then, the responses to the disasters were reactive:

more animal sacrifices, building bigger temples, developing new medicines, drilling deeper wells, seizing distant and more fertile lands, etc.

The worldview of Older Cultures rarely brought them to the inevitable and cyclic crises Younger Cultures have faced since their first eruption five thousand to seven thousand years ago. Because people in these Older Cultures assumed that humans were intrinsically good, emphasis was placed on nurturing and healing, rather than controlling and punishing. Because they believed that humans and natural systems were not separate but, instead, interpenetrated and interdependent—synergistic—they developed cultural, religious, and economic systems that preserved the abundance of their natural environment and provided for their descendants generation after generation.

So what are the easy answers to difficult problems?

Unlike many of our self-assured gurus, ecologists, and technologist something-will-save-us believers, I don't claim to know the exact details of our future. What I do know is that if we are to save some part of this world for our children and all other life, the answers won't simply rest in just the application of technology, economy, government, messianic figures, or new religions/sects/cults.

True and lasting solutions will require that a critical mass of people achieve an Older Culture way of viewing the world: the perspective that successfully and sustainably maintained human populations for hundreds of thousands of years.

Because I'm convinced that our problem is rooted in our worldview, the solutions offered in this book derive from ways we can change that, which will then naturally transform the technological/political/economic details that emerge from that new perspective. For example:

1.   History demonstrates that the deepest and most meaningful cultural/social/political changes began with individuals, not organizations, governments, or institutions.

2.   In helping to "save the world," the most important work you and I face is to help individuals transform their ability to perceive reality and control the stories they believe—because people do tend to live out what they believe is true. This has to do with people taking back personal spirituality, finding their own personal power, and realizing that most of our religious, political, and economic institutions are Younger Culture dominators and must be transformed if we are to prevent them from destroying us.

3.   Then, out of this new perspective, we ourselves will come up with the solutions in ways that you and I right now probably can't even imagine.

In the reality and experience of an Older Culture perspective, a life-connected worldview, we find a life rich and deep with wisdom, love, and the very real experience of the presence of the sacred in all things and all humans. A world that works for every living thing, including our children's children's children.

We Have Much to Learn . . . and Even More to Remember

 The Ancients knew something, which we seem to have forgotten.

—ALBERT EINSTEIN (1879–1955)

We started the journey of this book by plunging into the middle of a dense, dark forest. We learned about the situation we've gotten ourselves into, and the unavoidable evidence that we and our children are heading for challenging times. We've forgotten how to "live within our means," so our lives have become dependent on ancient sunlight (and other limited resources) stored in the ground, and the end of that supply is within sight. We're using up the Earth, and killing off our fellow living beings in the process.

From there we looked backward and learned how things got to be this way. We learned about the pivotal, world-changing importance of the stories we tell ourselves, the long and usually honorable history of our ancestors' sustainable Older Culture lifestyles—which most often didn't have this problem—and the shift, relatively recently, to Younger Cultures, the *wétiko* "dominator" city-state.

It won't be pleasant when the oil runs short. From Sumeria onward, every time a city-state's appetite for growth has run out of fuel, things have become unpleasant. But it's possible to survive, to make it through to whatever new life lies on the other side. The odds are that other sources of energy will emerge to replace much of what we get now from oil. The challenge, however, is that if the new energy sources are used in a Younger Culture way that merely allows for more humans to proliferate across our planet, destroying more competing species and resources and engaging in more wars of extermination against each other and the natural world, it may seal our doom even more solidly than would the loss of our fossil fuel resources.

Our energy sources are not as important as our view of life, which is grounded in our culture. This is what must change, and the need for that change is urgent.

The good news is that we don't have to invent a new culture or way of living. We do have much to learn, much to remember—the ways of our ancient ancestors, who lived sustainably long before we were born. Their way of life worked regardless of available energy stores, because its inherent connection to all other life built into it an extraordinary flexibility. That flexibility is still available to us, and even adaptable to our "modern" world.

The hundred thousand years of human history took about five thousand generations. Throughout almost all of that, most of our ancestors saw creation as sacred and treated both the natural world and one another with respect and reverence. Only in the last few hundred generations have we gone astray.

Among your own roots are thousands of generations of parents who survived sustainably, well enough to eventually give birth to you. You carry within you DNA descended from those very people. Imagine the chain of five thousand mothers who gave birth to your ancestors and eventually to you—leading all the way back to tribal life, and all the way forward to you.

Their ability to respect all life—to feel the presence of the divine intelligence in all living things and even in the supposedly inanimate universe—is as much within the genes your tribal ancestors gave you as is the instinct to create community and live together in cooperative harmony. The anthropological record shows us that psychologist Abraham Maslow was right when he hypothesized that human nature is good and instinctively seeks the divine, and that humans only become dysfunctional when they grow up in a sick culture that produces violent and damaged humans.

Hundreds of thousands of years of human history—and the modern-day "primitive" people we can still find alive on the Earth—tell us that the "conventional wisdom" that "man's innate nature is evil and dominating" is a lie, a sickness unique to our culture, and a relatively recent one in the long history of the human race. Instead, we are born to an innate knowledge and awe of the divine in all creation, and our first and most basic instincts are compassion and love.

In this moment, as you look around at the living world vibrating with life and vitality, energy that you can feel pouring from creation and into your heart as love, you—in this moment—connect with that ancient way of life and its sacred view of the world.

From that connection, that grounded place in the sacred here and now, you touch the power of life and transform yourself and—thus—those around you. They transform others and eventually every other living being on the planet. As you shift your view of the world—and thus begin to practice small, anonymous acts of mercy and compassion, change your ways of living and consuming, invest your rituals with spirit—your life will easily and naturally bring about the new ways of living that are necessary as we face the last hours of ancient sunlight.

Through this simple, practical, daily process, we begin to save the world.

Afterword

By Neale Donald Walsch,
author of *Conversations with God*

You have just read one of the most important books you will ever read in your life.

And because you have gotten this far in this extraordinary book, you are one of the Crucial Ones. You are one of the people who will play a key role in co-creating our future on this planet. You may not have thought of yourself in that role, but if you've gotten this far in this book, you've been given it.

That's how Life works. That's how the Universe functions. That's how God converses with all of us. First we are confronted with data, information—a communication. Then we are invited, urged, or compelled to absorb the information, to receive the communication. Finally, we decide Who We Are in relationship to it.

That's what you're doing now. You're deciding Who You Really Are in relationship to the incredibly important information you've just absorbed. And now, no matter what you decide, you will play a key role in co-creating the future on Earth.

If you decide to ignore this information, you will co-create one kind of future. If you decide to act on it, you will co-create another.

You can't step out of your role now. You know too much.

When I first read this book I knew that I could never view my life in the same way again. I could see myself as part of the problem, or as part of the solution, but I could never again see myself as having nothing to do with either.

Somewhere in the middle of this book you may have been saying to yourself, "I see the problem. I get it! But what can I do?" Hopefully now, with you having completed the book, that question has been answered. But there is another question that must immediately follow for all thinking people: "Can it work?"

I'm here to tell you that it can. But much will depend—everything will depend—on whether you believe that it can, know that it can, intend that it will.

We are now engaged in the process of what Barbara Marx Hubbard, author of *Conscious Evolution,* calls conscious evolution. We are re-creating ourselves anew on this planet, with every daily decision and choice we make in every Moment of Now, and we are doing so—perhaps for the first time in human history (and especially after reading this book) with full awareness of what we are creating, and how.

I am encouraging you with all my heart not to put this book down unresolved. Thom Hartmann has presented here a plan of action. He has given you tools you can use, beginning right here, right now, in helping to change the collective consciousness, and write a new "story" with which to fuel the engine of the human experience.

If you do not think that one person can do much—can do enough—to make a real difference, I urge you to read another of Thom's books, *The Prophet's Way.*

Get it now. Read it immediately. It will inspire and excite you. For it will show you in real-life terms just what one person can do, and lift you to a place of new determination to play your rightful role in the Creation of Tomorrow.

As for this book, quote it everywhere. Buy 10 copies and give them away. Don't let this call to action go unheard.

You may feel like a voice in the wilderness, but it is your voice we are waiting to hear. Yours is the crucial vote. You are the determining factor. We reach Critical Mass when we reach you—and you choose to reach others—with the simple message of this book: We are all One.

Let us act, at last, in the best interests of us all. Then the Sun will shine another day, and another still, and life will not merely go on, but achieve its highest expression, its grandest glory, its greatest joy. Can we give this gift to our children?

Please say yes.

Neale Donald Walsch
Ashland, Oregon

Notes

Introduction

1. www.wri.org/wr2000/page.html

We're Made Out of Sunlight

1. The exception to this is the bacteria and living organisms on the ocean's floor, miles below the surface, which live off the heat of undersea volcanic vents. Even these, though, are living off energy from a sun; the core's volcanic heat was stored when the Earth was first formed from the exploding core of a star/sun.

2. While coal is clearly ancient vegetation, there is a debate about the origin of oil. Conventional wisdom holds that it's also vegetative in Nature, probably from sea vegetation, but another theory put forth by Cornell University astronomy professor Thomas Gold holds that oil is created by hyperthermophilic (high-temperature-living) bacteria at depths ranging from 8 to 100 kilometers below the surface. While Gold's theory—which has intriguing supporting evidence, such as the presence of helium in natural gas—would mean that oil is not "ancient sunlight," it does not alter the central thesis of this book, since the process of bacterially producing today's usable oil by Gold's proposed means is also a multimillion-year-long process. Once current stores are exhausted, it will take hundreds of millions of years to replenish them. (See "The Deep, Hot Biosphere" by T. Gold, *Proceedings of the National Academy of Sciences,* 89:6045–49.)

3. *World War III* by Michael Tobias, Bear & Co., 1994.

4. In Petroconsultants' study, "The World Oil Supply 1930–2050."

5. *The Golden Century of Oil: 1950–2050: The Depletion of a Resource* by C. J. Colin Campbell, Kluwer Academic Pub., Norwell, Mass.

How Can Things Look So Good Yet Be So Bad?

1. *The Heat Is On* by Ross Gelbspan, Addison-Wesley, 1997.

2. *Public Health Reports,* U.S. Department of Health and Human Services, January–February 1996, vol. 111, no. 1, p. 8(2).

3. "Drug-Resistant TB May Bring Epidemic" by Barbara J. Culliton, *Nature,* September 1992, vol. 356, no. 6369, p. 473(1).

4. "The Third Epidemic-Multidrug-Resistant Tuberculosis," *Chest,* January 1994, vol. 105, no. 1, p. 32.

5. *And the Waters Turned to Blood* by Rodney Barker, Simon & Schuster, 1997.

6. *Deadly Feasts* by Pulitzer Prize–winning author Richard Rhodes (Simon & Schuster, 1997) is an excellent book on the topic.

Glimpsing a Possible Future in Haiti and Other Hot Spots

1. Letter of Columbus quoted in *Documents of West Indian History* by Eric Williams, Port-of-Spain, Trinidad: PNM, 1963, and Peter Martyr's, *De Orbe Novo,* 1516.

The Death of the Trees

1. *Diet for a New America* by John Robbins, H.J. Kramer, 1987.
2. Dirk Beveridge, Associated Press, August 25, 1997.

Extinctions: Diversity Supports Survival

1. *The Sixth Extinction: Patterns of Life and the Future of Humankind* by Richard Leakey and Roger Lewin, Anchor Books, 1996.
2. The word "huge" is not excessive: food production is a far bigger business than most people realize, is tightly controlled, and has little competition. Two companies, for example, Cargill and Continental, controlled 50 percent of all U.S. grain exports in 1994, a year when U.S. grains constituted 36 percent of wheat; 64 percent of corn, barley, sorghum, and oats; and 40 percent of soybean exports worldwide.
3. *Beyond Growth* by Herman Daly, Beacon Press, 1997.

Climate Changes

1. "Global Warming Is Marmot Wake-Up Call," *Science News,* April 29, 2000, vol. 157, p. 282.
2. Apparently some carbon dioxide has also been removed from the atmosphere by the Himalayan mountains, which expose upper-atmosphere air to rock and form carbonate forms of some of those minerals.
3. *The End of Nature* by Bill McKibben, Anchor Books, 1999.
4. *The Dying of the Trees* by Charles Little, Penguin Books, 1997.
5. www.aip.org/pt/vol-55/iss-8/captions/p30cap2.html.
6. "Here Comes the Rain," by Nicolas Jones, *New Scientist,* April 26, 2003, pp. 24–25.
7. *The Time Before History* by Colin Tudge,·Touchstone Books, 1997.
8. *A Brain for All Seasons* by William H. Calvin, University of Chicago Press, 2002.

A Visit to a Country That's Planning How to Survive: China

1. I've changed the details and the name of the doctor to protect his safety.
2. *Who Will Feed China?* by Lester Brown, W.W. Norton, 1995.

The Last Hours of (Cheap, Clean) Water

1. wmc.ar.nrcs.usda.gov/tech.dir/droughtmgmt.htm.
2. www.mvm.usace.army.mil/grandprairie/area/default.asp.
3. www.ficus.usf.edu/docs/injection_well/sutherland1.htm.
4. "Drugged Waters: Does It Matter That Pharmaceuticals Are Turning Up in Water Supplies?" by Janet Raloff, *Science News,* March 21, 1998.

5. "Frogs Rapidly Vanishing Across U.S., Experts Unsure of the Cause" by Traci Wilson, Gannet News Service, *Burlington Free Press*, August 16, 1998.
6. *The Lost Language of Plants* by Stephen Harrod Buhner, Chelsea Green, 2002.

Deforesting, Fighting for Fuel, and the Rise and Fall of Empires

1. archive.greenpeace.org/pressreleases/climate/1998jun9.html.
2. Some people, hearing about this, say, "That war wasn't about oil; our ally, Kuwait, got invaded by the bad guys." But our allies get attacked and invaded all the time, somewhere in the world. America's response, in the late twentieth century, has only been this ferocious when oil is at stake.
3. In the world of international business, a one-month or one-quarter profit-and-loss statement has become far more important than projections of 40-year sustainability, as we've seen in industry after industry. And corporations, in many cases, have replaced nations as centers of wealth and power, becoming the "new dominators." While Indonesia is the twenty-third largest economy in the world, it's smaller than Mitsubishi, which is twenty-second. Denmark and Thailand are dwarfed by number 26, General Motors. And Exxon, Hitachi, Toyota, AT&T, and Shell are all in the top 50. Corporations, increasingly, have the power to manipulate public opinion and force elected officials to bow to their will in every nation of the world.
4. Sunlight powers the water cycle by evaporating water, causing it to rain back down into mountain reservoirs. It also provides the heat energy that makes the winds circulate for wind power.

Younger Culture Drugs of Control

1. For an exhaustive and shocking insight into this problem, read *The Media Monopoly* by Ben Bagdikian, Beacon Press, Boston and *The FAIR Reader* by Jim Naureckas and Janine Jackson Westview Press/HarperCollins, New York.
2. For details, call or write Food & Water, Walden, Vermont, 1-800-EAT-SAFE. June 22, 1997, *Burlington Free Press*.
3. *Civilization and Its Discontents*, by Sigmund Freud, various editions.
4. *The Voice of the Earth* by Theodore Roszak, Phanes Press, 2002.
   Ecopsychology by Theodore Roszak, Sierra Club Books, 1995.
5. *The Origin of Consciousness in the Breakdown of the Bicameral Mind* by Julian Jaynes, Houghton Mifflin, 2000.
6. *Food of the Gods* by Terence McKenna, Bantam Books, 1993.

Younger Culture Stories About How Things Are

1. *Valuing the Self: What We Can Learn from Other Cultures* by Dorothy Lee, Prentice Hall, 1958.
2. *Columbus and Other Cannibals* by Dr. Jack Forbes, Autonomedia, 1992.
3. I recently learned that while some Native American tribes have had written languages for thousands of years, others have resisted writing down their language. The Apache language, for example, was first written down and codified only three decades ago by a Methodist missionary; an Apache told me, "It was a mistake to do

that: our language is too sacred to be written down." It would be interesting to explore the differences in the nature of the personal religious experiences of the written- and non-written-language peoples. To the best of my knowledge, nobody has done such an exploration.

4. www.stopnato.org.uk/du-watch/caldicott/medico.htm.

5. *America: What Went Wrong* by Donald L. Barlett and James B. Steele, Andrews McMeel Publishing, 1992.

6. *The Chalice and the Blade* by Riane Eisler, HarperSanFrancisco, 1987.

What We Need to Remember

1. Although it would still be over 100 years before Jennings would develop a vaccine for smallpox, nevertheless Europeans had been exposed to it for centuries, perhaps millennia. This gave them a relative immunity to it: those with genes that were vulnerable to it had mostly died, long ago. So, usually only 5 percent to 30 percent of unvaccinated Spanish would die from the disease. The Incan population, however, never having been exposed to it, is estimated by some authorities to have had death rates between 60 percent and 95 percent following their first exposure in 1520.

2. *The Great Forgetting* by Geoff Page and Bevan Hayward, published in Australia, ISBN 0-85575-290-4.

3. *Ishmael* by Daniel Quinn, Bantam Books, 1995.
   The Story of B by Daniel Quinn, Bantam Books, 1997.

4. Which is why Pol Pot marked for extermination first those who wore glasses and the elderly.

5. Early Europeans had considerable fear of Native American shamanic powers, which included the claimed ability to make rain and otherwise control the natural world. The first English colony founded in the United States was in 1587–1588 on Roanoke Island, North Carolina, and featured the first murder of a Native American chief by whites and the first birth of a European on American soil (Virginia Dare). When the British came back to the colony in 1589 to bring provisions and supplies, they discovered that all four hundred of the colonists had vanished without a trace, causing this to be referred to as "The Lost Colony." The second English attempt at colonization occurred in Jamestown, Virginia, in 1607: only 38 of the original 104 settlers survived the first year, and another 4,800 colonists starved in subsequent attempts to colonize the area over the following seven years. But why? Matthew Therrell, a tree-ring specialist with the University of Arkansas, studied rings of recently cut thousand-year-old bald cypress trees and found a startling anomaly, published in the April 24, 1998, issue of the journal *Science*. Between the years 1000 and 1997, there were two—and only two—massive and tree-withering droughts along the East Coast . . . during the years 1587–1588 and 1607–1614.

6. The church also had internal power struggles about these lands. In 1997 Louise and I visited the Gila River Pima Indian Community. Old maps from the 1700s show the area as once being "Franciscan lands," then later "Jesuit lands," after the earliest Spanish invaders had left with all the gold.

7. *Health and the Rise of Civilization* by Mark Nathan Cohen, Yale University Press, 1989.

8. While the earliest evangelical cultures are lost in history, we can see the internal debates of this in the writings of the ancient Greeks (who did not evangelize) and the Romans. The Romans originally did not evangelize, as can be seen in the writings of a number of their philosophers and leaders, including Julius Caesar. Roman citizenship was an "exclusive club." But as the empire was crumbling in the early fourth century, Emperor Constantine made a bid to save his empire by officially adopting the Jewish Messiah Jesus to replace the sun-god the Romans had been worshipping, and changed the Jewish Sabbath (Saturday) day of worship to the day that the Romans had traditionally worshipped the sun-god (Sunday). With this stroke, creating the one ("Catholic") official Roman church, Constantine assimilated into Roman culture the evangelistic concepts most clearly articulated in the writings of Paul. While this is by no means the only example of a culture adopting evangelism, it's one of the most well documented.

9. "Behold the fowls of the air: for they sow not, neither do they reap, nor gather into barns; yet your heavenly Father feedeth them." (Jesus, in Matthew 6:26)

10. *Man's Rise to Civilization* by Peter Farb, Penguin Books, 1992.

The Lives of Ancient People

1. *The Heart of the Hunter* by Laurens van der Post, Penguin Books, 1961.

Power vs. Corporation in Social Structure: the City-State vs. Tribes

1. *Man's Rise to Civilization: The Cultural Ascent of the Indians of North America* by Peter Farb, Bantam Books, 1978.

2. *The Prehistory of the Mind: The Cognitive Origins of Art, Religion and Science* by Steven Mithen, Thames & Hudson, 1999.

3. *Victims of Progress* by John Bodley, Mayfield, 1990.

4. "The Diagnosis and Treatment of Athletic Amenorrhea" by Susan L. Epp, *Physician Assistant,* March 1997, vol. 21, no. 3, p. 129(9).

5. "Ovarian Function and Disease Risk" by C. La Vecchia, *Cancer Researcher Weekly,* October 25, 1993, p. 17 (1).

6. *Living the Spirit: A Gay American Indian Anthology* by Will Roscoe, St. Martin's Press, 1988.
Spirit and the Flesh: Sexual Diversity in American Indian Culture by Walter Williams, Beacon Press, 1992.

The Robots Take Over

1. *Grunch of Giants* by Buckminster Fuller, www.bfi.org.

The New Science

1. *Holographic Universe* by Michael Talbot, HarperCollins, 1991.
2. *Web Without a Weaver* by Victor Grey, Open Heart Press, 1997.

New Stories Are Necessary to Change the World

1. *Virus of the Mind* by Richard Brodie, Integral Press, 1995.

Touching the Sacred

1. This is an original translation from the original Spanish, done for this book by my dear friend Lilio Aragones, for which I am most grateful.

Learn to Create Awareness

1. There is more information about the Kogi in my book *The Prophet's Way* and also in an excellent videotape available from Mystic Fire Video called *From the Heart of the World* and *The Elder Brothers* by Alan Ereira, Knopf, 1992.

Respect Other Cultures and Communities

1. As whites figured this out, it became a common U.S. government practice to select a single tribal person—often one who could be corrupted with money or alcohol—and designate that person as the sole authority on behalf of the tribe. Then "treaties" could be executed with the appearance of legality, and when the rest of the tribe resisted giving up their lands, minerals, or whatever was in the treaty, they could be brutally repressed for "treaty violations." This practice continues to this day under the Bureau of Indian Affairs and other U.S. government agencies and large corporations who want Native lands.

2. *Beyond Growth* by Herman Daly, Beacon Press, 1997.

3. When Ronald Reagan came to power in 1980, the top U.S. income tax bracket was 70 percent—similar to most of Europe. When he left office, the top bracket was 28 percent, bringing a surge of wealth to the richest Americans, and generating a $3-trillion national debt.

4. Cited in *Unequal Protection: The Rise of Corporate Dominance and the Theft of Human Rights* by Thom Hartmann, Rodale Press, 2002.

5. Where the Marxists/Communists went wrong in this was that they measured "wealth" as goods, services, and capital, so that's what they tried to control. They failed to understand that in Older Cultures "wealth" meant safety, security, and the daily experience of the sacred: the roots of human needs. These needs (and particularly the need to touch the sacred) are very, very difficult to meet with a social group as large as a nation-state. "To each according to his needs; from each according to his abilities" is a very basic Older Culture concept, but the Communists applied it in a Younger Culture context, so it was doomed to fail from the beginning. In trying to redistribute material wealth, they simply created a new class of Younger Culture dominator wealth-controllers: the bureaucrats.

Renounce the Destruction of Life

1. *Ishmael* by Daniel Quinn, Bantam Books, 1995.

Turn Off the TV

1. "Kids Who Watch TV More Likely to Drink" by Eric Fidler, Associated Press, date unavailable (est. 2000).

The Modern-Day Tribe: Intentional Community

1. *Famine and Survival in America* by Howard Ruff, Ruff Times Press, 1974.
2. *Creating Community Anywhere* by Carolyn Shaffer, et al., Perigee, 1993.
   Communities magazine published by the Fellowship for International Community, www.ic.org.
   Communities Directory written and published by Fellowship for Intentional Community, 2000.
3. *Spirituality and Community* by Don Calhoun, self-published and now out of print.

Reinventing Our Daily Life and Rituals

1. *The Prophet's Way* by Thom Hartmann, Three Rivers Press, 1999.

Transforming Culture Through Politics

1. *The Praetorian Guard: The U.S. Role in the New World Order* by John Stockwell, South End Press, 1990.

Taking Back America

1. *The Emerging Republican Majority* by Kevin Phillips, Arlington House, 1969.
   Blinded by the Right by David Brock, Three Rivers Press, 2003.
2. *How Democratic Is the American Constitution?* by Ronald A. Dahl, Yale University Press, 2002.

Recommended Reading

Many people have written to ask for suggestions of where they could find more information on the ideas presented in this book. This list is by no means complete, but it's a start. We'll be updating it in future editions of the book and on our website as time goes on.

And this list is not a bibliography: many of the books referenced here are listed in the Notes, as we chose to cite bibliographic content in context. This list, instead, is a somewhat eclectic collection of books that I found meaningful or useful, as well as relevant to the topic of this book, in addition to those books cited throughout the text.

You may also want to check my website at www.thomhartmann. com for other recommendations and essays and updates on the topics of this book.

And the Waters Turned to Blood by Rodney Barker
Asian Journal, The by Thomas Merton
Bare-Bones Meditation by Joan Tollifson
Be Here Now by Ram Dass
Beyond Growth by Herman E. Daly
Black Elk by Wallace Black Elk and William Lyon
Breakout by Marc Lappé
Case Against the Global Economy, The edited by Jerry Mander and
    Edward Goldsmith
Chalice and the Blade, The by Riane Eisler
Columbus and Other Cannibals by Jack Forbes
Coming Plague, The by Laurie Garrett
Communities (magazine) published by Communities Magazine,
    Alpha Farm, Deadwood, OR 97430
Communities Directory, The published by the Fellowship for
    Intentional Community, PO Box 814, Langley, WA 98260

Conscious Evolution by Barbara Marx Hubbard

Creating Communities Anywhere by Carolyn Shaffer and Kristin Anundsen

Dark Night of the Soul by St. John of the Cross

Dawn Land by Joseph Bruchac

Deadly Feasts by Richard Rhodes

Deep Breath of Life, A by Alan Cohen

Diet for a New America by John Robbins

Dying of the Trees, The by Charles Little

End of Nature, The by Bill McKibben

End of Work, The by Jeremy Rifkin

Evolution's End by Joseph Chilton Pearce

Food of the Gods by Terence McKenna

Forest People, The by Colin M. Turnbull

Great Work, The by Thomas Berry

Guns, Germs, and Steel by Jared Diamond

Harmless People, The by Elizabeth Marshall Thomas

Healing Ceremonies by Carl A. Hammerschlag and Howard D. Silverman

Healing of America, The by Marianne Williamson

Heart of the Hunter, The by Laurens van der Post

Heat Is On, The by Ross Gelbspan

Holographic Universe, The by Michael Talbot

Honest to Jesus by Robert W. Funk

Hopi Survival Kit by Thomas Mails

Humanity's Descent by Rick Potts

In the Absence of the Sacred by Jerry Mander

Ishmael by Daniel Quinn

Kalahari Hunter-Gatherers by Richard B. Lee and Irven DeVore

Lies My Teacher Told Me by James W. Loewen

Limited Wants, Unlimited Means edited by John Gowdy

Man's Rise to Civilization by Peter Farb

Markings by Dag Hammarskjöld

Media Monopoly, The by Ben Bagdikian

My Ishmael by Daniel Quinn

Mythic Life, A by Jean Houston

Population Explosion, The by Paul and Anne Ehrlich
Powers That Be, The by Walter Wink
Presence of the Past, The by Rupert Sheldrake
Reworking Success by Robert Theobald
Ruben Snake, Your Humble Serpent by Jay Fikes
Saving the Planet by Lester R. Brown, Christopher Flavin, and
    Sandra Postel
Self-Aware Universe, The by Amit Goswami
Shaman's Doorway, The by Stephen Larsen
Sixth Extinction, The by Richard Leakey
Stone Age Economics by Marshall David Sahlins
Technopoly by Neil Postman
Time Before History, The by Colin Tudge
Tribes by Art Wolfe
True Hallucinations by Terence McKenna
Unnatural Order, An by Jim Mason
Voice of the Earth, The by Theodore Roszak
Voices of the First Day by Robert Lawlor
When Corporations Rule the World by David C. Korten
Who Will Feed China? by Lester R. Brown
World War III by Michael Tobias

Acknowledgments

While most of the ideas in this book have been rolling around in my head for years (as readers of my earlier books will recognize), the works of several people have helped distill them more clearly for me in the past few years. The most prominent of them is Professor Jack D. Forbes of the University of California at Davis, who was kind enough to endure my visit and many e-mails and encourage my work. Others, who have transformed my world at a bit more of a distance through their writings, include Ross Gelbspan, Bill McKibben, Charles Little, Joseph Chilton Pearce, Dan Millman, Daniel Quinn, Riane Eisler, Lester R. Brown, Paul and Anne Ehrlich, Michael Tobias, Rupert Sheldrake, Jerry Mander, Richard Bandler, Alan Cohen, Patch Adams, Theodore Roszak, John Robbins, Terence McKenna, James Lee Burke, and Jack Vance. I owe them all a large debt for the work and thought they gave to their books and publications.

My editors, proofreaders, and helpers on the first edition of this book—Dave deBronkart, Kyle Roderick, Brad Walrod, Jerome Lipani, and Gwynne Fisher—have provided invaluable input. And my editor on the second edition, Toinette Lippe, was absolutely brilliant, helping reshape it into a book well suited for the twenty-first century.

Thanks to Hal and Shelley Cohen for being good sounding boards and friends, to Jack and Norma Vance for their encouragement and example of what a true commitment to writing means, and to Rita Curtis, Tammy Nye, Adam Cohen, Rob Kall, Karen Cross, Ellen Lafferty, Susan Reich, Tim Underwood, and Charlie Winton. I am profoundly grateful to editor Peter Guzzardi at Harmony Books, who cared enough to bring this work to a larger public, and to my agents, Bill Gladstone and Stephen Corrick, who worked so hard to make its publication possible.

The most important of all the influences that have led to this book have been my parents, Carl and Jean Hartmann; my wife, Louise Hartmann; my children, Kindra, Justin, and Kerith Hartmann; and my mentor for many years, Gottfried Müller: to them I owe not only my life but the quality of my aliveness, for which I am deeply grateful.

Finally, I give a very special thanks to the founders and staff of Worldwatch Institute, Cultural Survival, and EarthSave for the critically important work they are doing in chronicling the state of the planet and offering practical solutions for change.

Index

About the Author

THOM HARTMANN's books have been featured in *Time* and on NPR's *All Things Considered*, CNN, and the front page of the *Wall Street Journal* (twice). Over the past two decades he has spoken to more than 100,000 people on five continents and is heard daily by people from coast to coast in the U.S. and around the world on *The Thom Hartmann Radio Program*. He lives in central Vermont. Visit him at www.thomhartmann.com.

Also by Thom Hartmann

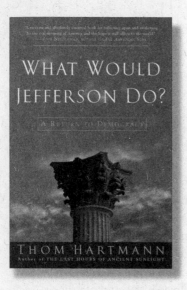

WHAT WOULD JEFFERSON DO?

A Return to Democracy

1-4000-5209-2
$14.00 paperback
(Canada: $21.00)

First tried in Athens in 508 B.C. and then not again until it was embraced by the Founding Fathers in 1776, democracy is now at a crossroads, assailed by powerful forces that seek to corrupt it. Thom Hartmann offers a blueprint for revitalizing and reforming that which we hold most dear.

 THREE RIVERS PRESS

On sale wherever books are sold
www.crownpublishing.com